THE AMBIGUOUS FROG

Marcello Pera

THE AMBIGUOUS FROG

THE GALVANI-VOLTA CONTROVERSY ON ANIMAL ELECTRICITY

Translated by Jonathan Mandelbaum

PRINCETON UNIVERSITY PRESS   PRINCETON, NEW JERSEY

Library of Congress Cataloging-in-Publication Data
Pera, Marcello, 1943–
[Rana ambigua. English]
The ambiguous frog : the Galvani-Volta controversy on
animal electricity/ Marcello Pera ;
translated by Jonathan Mandelbaum
p.   cm.
Translation of: La rana ambigua.
Includes bibliographical references and index.
ISBN 0-691-08512-9 (alk. paper)
1. Electrophysiology—History.   2. Galvani, Luigi, 1737–1798.
3. Volta, Alessandro, 1745–1827.   I. Title.
QP341.P4713   1992
591.19'127—dc20   91-3077

To Antonia

———————————————

Contents

List of Illustrations

Note: See captions for source of each figure. Unattributed illustrations are by the author.

UNTIL RECENTLY the early history of electricity has remained the almost exclusive province of historians of science, never becoming a primary topic of interest to philosophers to the degree that occurred in mechanics, heat, atomism, and other branches of physics. Now that situation has radically changed. The keen philosophical discernment of Thomas S. Kuhn has caused us to behold in a radically new way the diagrams in Volta's dramatic announcement of the electric battery (or Voltaic cell) in 1800 (Kuhn 1987). In the work before us, Marcello Pera shows how a fresh examination of the Galvani-Volta controversy enables us to have a new perspective on the problems of choosing between competing theories.

The science of electricity was born in the eighteenth century. Although a few phenomena, such as the attraction and subsequent repulsion of small bits of matter by various rubbed objects, had been known for centuries, there was no coordinated knowledge adequate for the establishment of a new science until about the 1740s. For example, Isaac Newton wrote about some aspects of electric attraction and repulsion but he had no idea that some substance were conductors of electricity while others were nonconductors. Nor was any scientist in Newton's day aware of the simple fact that there are two modes of electrification, one associated with rubbed glass and the other with rubbed amber. By the mid-eighteenth century the catalog of electrical phenomena had grown by leaps and bounds. The most spectacular innovation was the Leyden jar, the first condenser or capacitor, which apparently had the power to accumulate vast quantities of "electricity," which it could release suddenly with great "force." By mid-century, scientists were well aware of the facts of conduction and insulation, and even the effects of grounding in electrostatic experiments.

Electricity achieved the full status of a science in the 1740s, when Benjamin Franklin put forth a theory of electrical action based on the motion or transfer of a single electric "fluid," in producing a "positive" or "plus" state or charge if in excess, and a "negative" charge in default. The spectacular electrical experiments designed by Franklin—the experiment of the sentry box and the lightning kite—not only tested his hypothesis that the lightning discharge is an electrical phenomenon, but showed that electrical phenomena arise without human artifice in the world of nature.

All of these developments were part of that branch of electricity known as electrostatics. Professor Marcello Pera is concerned with the birth of a new branch of electricity, current electricity, originating at the end of the eighteenth century. It is well known that the first stages of this new branch of electrical science are associated with Luigi Galvani, a professor of anatomy and of obstetrics at the University of Bologna. His seminal paper described experiments he had performed with dissected limbs of frogs in the presence of an electrostatic generator. Galvani was well acquainted with Franklin's one-fluid theory of electricity and attempted to explain his results by recourse to a similar kind of "fluid" of animal electricity. Allessandro Volta repeated these experiments and at first followed Galvani in using the hypothesis of an animal electricity. Before long, however, Volta had transformed the subject by a series of delicate experiments made possible in part by his invention of a very sensitive detecting device that he called the "condensing electroscope."

In time Volta abandoned Galvani's concept of an animal electricity and produced his own theory, based on the contact of different metallic or conducting substances. Volta's work reached a climax when it led him to an invention that has had momentous repercussions for science and for human society. This invention was the electric battery and the possibility of producing a continuous flow of electricity or an electric current. Volta's announcement of the electric battery in 1800 inaugurated the electric age.

Marcello Pera has based his presentation on extensive reading in the primary sources. He has not, however, produced a conventional recasting of a familiar story. Rather, approaching the subject with the keen analytic tools of the philosopher, Professor Pera has presented the conflict between Galvani and Volta in a wholly new light. Traditionally, an axiom of science has held that when there are two rival theories a crucial experiment will decide between them. In the case of the rival theories of Volta and Galvani, however, Professor Pera shows that there was never a single experiment produced either by Volta or by Galvani and his followers that uniquely decided in favor of one or the other.

This conclusion is of more than ordinary interest. It not only presents the contest between Galvani and Volta in a new light; it also raises a question of great interest to philosophers and historians of science, and to scientists as well: What are the reasons why one theory succeeds and another fails? So long as both theories could produce explanations of the phenomena, why was Volta victorious? Why did the galvanic system ultimately fail?

The conflict between Galvani and Volta seems to have been resolved by the force of circumstances other than experimental evidence for one rather than the other. One factor of some importance was the early death (1798) of Galvani and the assumption by his nephew Aldini of the position of primary advocate. Aldini was no match in scientific debate for Volta. Not only did Aldini lack the scientific greatness of his rival, but he shifted the sober discussion of the new science from the laboratory, the academy, and the pages of scientific journals to the arena of public entertainment, performing bizarre spectacles as a professional showman. That is, Aldini became a mountebank, demonstrating the effects of his uncle's discovery by electrically animating the head of a calf severed from its body, causing the eyes, tongue, and mouth to move or twitch. His most spectacular performance was to produce the same effect in the body of a criminal recently dead from a public hanging. For the world of sober science, Aldini had in effect given over the laurel to his uncle's rival.

Of even greater significance was Volta's invention of the electric battery. This invention—which took two forms, the "pile" and the "crown of cups"—was associated with Volta's theory of contact electricity. The battery not only transformed the subject of electricity from electrostatic to current electricity, but affected other branches of science such as chemistry and biology. Before long the arc light showed that the voltaic battery could have enormous significance for technology, even before the discovery of electromagnetism made possible the production of electric motors and generators.

The astonishing effects of Volta's invention sealed his fame and guaranteed the success of his theory over Galvani's. As Professor Pera shows, however, this complete victory of Volta's did not hinge on any specific feature that—at least in the early days—Galvani could not explain. Readers will rejoice that their guide to this episode is Marcello Pera, a scholar distinguished for historical learning coupled with philosophical acumen. His presentation gives us reason to hope anew for the fruitful results that will come from the happy bonding of secure knowledge of the history of science and true philosophic insight.

I. Bernard Cohen

Acknowledgments

FOR THE general historical background, a work such as this follows in the foot-steps of giants who well deserve our grateful tribute. No one who approaches the early history of electricity can dispense with contributions as monumental as those of Mario Gliozzi, I. Bernard Cohen, and John Heilbron. Although I have always based my own work on original sources, it is only proper to acknowledge my considerable debt to these scholars, even when I have departed from their findings.

As for the section directly concerning Galvani and Volta, my gratitude goes to the editors of the seven volumes of the *Edizione nazionale* of Volta's works and the five volumes of his correspondence; to Angelo Ferretti-Torricelli, the masterful compiler of the *Indici delle opere e dell'epistolario di Alessandro Volta* (2 vols., Milan: Rusconi Editore, 1974); and to Gustavo Barbensi and Mario Gliozzi, editors of Galvani and Volta's *Opere scelte* (selected works), respectively. The lack of a complete critical edition of Galvani's writings is a very serious gap in the historiography of Italian science.

On the Galvani–Volta controversy, another, special acknowledgment is due to Giovanni Polvani. His contribution to the *Edizione nazionale* and his pioneering monograph on Volta are so valuable that they may have discouraged others from undertaking similar work, although I find his reconstruction of the controversy too favorable to Volta and lacking in interpretative force. The reconstructions by Giulio Cesare Pupilli and Ettore Fadiga, and by Bern Dibner, seem to me more balanced, although not as analytical as Polvani's. The account of the dispute by Emil du Bois-Reymond remains essential.

But these are, precisely, value judgments. It is therefore right that I should assume responsibility for them and for my own interpretation of the controversy, which often differs from those of the authors cited above.

My historical reconstruction and epistemological interpretation have benefited from many suggestions, comments, objections and source leads. In particular, I am grateful to my friends Francesco Barone, Giulio Giorello, Maurizio Mamiani, Marco Mondadori, and Giorgio Tabarroni, who were the first to discuss my work with me while it was still in progress. The manuscript was already completed in 1984, when I had the pleasure of being invited as Visiting Fellow at the University of Pittsburgh's Center for Philosophy of Science. In that exciting intellectual atmosphere, I had the opportunity to discuss my views on the philosophical issues in the Galvani–Volta controversy with many friends who unsparingly offered useful comments. My personal affection is compounded by a special gratefulness to Adolf Grünbaum, Carl Hempel, Peter Machamer, Nick Rescher, and Merrilee and Wes Salmon. Aristides Baltas spurred me on for all these years with his loyal friendship and penetrating insights. Even before the Italian edition was published, my frog—escorted by me or entrusted to my writings—croaked in many parts of the world, receiving highly beneficial attention from many friends. I should like to give particular thanks to Paul Feyerabend for his comments; he will find in the closing section of the book a partial compensation for my criticisms of him elsewhere. I am also grateful to the late Bern Dibner, to Maurice Finocchiaro, Roderick Home, Larry and Rachel Laudan, Tom Nickles, Joe Pitt, and William Shea.

This American edition includes a new section (§6.5), in which I have tried to focus on the challenge posed by the Galvani-Volta controversy to the methodologist intent on rebuilding—or craving to dictate—the rules of scientific research. I have also consulted the new studies that have appeared in the meantime, notably the excellent paper by Naum Kipnis—with which my book has significant affinities—and a paper by Thomas Kuhn on scientific revolutions in which the change from the contact theory to the chemical theory of the Voltaic pile is examined. I am further indebted to Kuhn for discussing with me, in a stimulating correspondence, his view about this change and my interpretation of the controversy. John Heilbron provided precious encouragement and help, and I am very pleased to thank I. Bernard Cohen for his confidence in my work and for offering to write the foreword to the American edition. I thank Pietro Corsi for his generous assistance in preparing this volume and Jonathan Mandelbaum not only for his scrupulous translation but for his attentive revisions. Antonia—who fortunately knows nothing about frogs but understands the problems of those besotted with them—has, here again, proved irreplaceable.

Everyone mentioned above has seen and discussed with me one or more parts of my work, problems of historical interpretation, or issues in philosophical analysis. But not even an ambiguous frog could be so ambiguous as to let its audience construe an admission of assistance as a call for complicity.

TRANSLATOR'S NOTE

This translation was largely funded by a generous grant from the Euryalus Foundation, Siracusa, Italy.

Passages from Margaret Glover Foley's 1953 translation of Galvani's *Commentary on the Effects of Electricity on Muscular Motion* (GF: see List of Abbreviations) are reproduced with the kind permission of the Burndy Library, Norwalk, Connecticut, and Margaret G. Foley Ames.

Translations of other sources, particularly the writings of Galvani and Volta, are mine unless the references accompanying the quoted matter included a previous English version. In a few instances, I have adapted an existing translation to improve accuracy or readability, but such changes are always indicated in parentheses.

J.M.

List of Abbreviations

Sources are identified by the author-date system. The References at the end of the book are therefore limited to items quoted or mentioned in the text. For the sake of brevity and convenience, I have used the following abbreviations for Galvani's and Volta's works:

GF Luigi Galvani, *Commentary on the Effects of Electricity on Muscular Motion*, translated into English by Margaret Glover Foley, with notes and a critical introduction by I. Bernard Cohen, together with a facsimile of Galvani's *De viribus electricitatis in motu musculari commentarius* (1791) and a bibliography of the editions and translations of Galvani's book prepared by John Farquhar Fulton and Madeline E. Stanton (Norwalk, Connecticut: Burndy Library, 1953. Reprinted with permission of the Burndy Library and Margaret G. Foley Ames) Referred to in the text as *De viribus electricitatis* or *Commentarius*.

CM *Memorie ed esperimenti inediti di Luigi Galvani con la iconografia di lui e un saggio di bibliografia degli scritti* (Bologna: Cappelli, 1937). Contains the first complete Italian translation, by Enrico Benassi, of *De viribus electricitatis* and the notes thereon by Galvani's nephew, Giovanni Aldini.

GO *Opere edite e inedite del Professore Luigi Galvani*, raccolte e pubblicate per cura dell' Accademia delle Scienze dell'Istituto di Bologna (Bologna: Tipografia di Emilio Dall'Olmo, 1841).

GOS L. Galvani, *Opere scelte*, ed. G. Barbensi (Turin: Utet, 1967).

VE *Epistolario di Alessandro Volta*, "Edizione nazionale," 5 vols. (Bologna: Zanichelli, 1949–55).

VO *Le opere di Alessandro Volta*, "Edizione nazionale," 7 vols. (Milan: Hoepli, 1918–29).

VOS A. Volta, *Opere scelte*, ed. M. Gliozzi (Turin: Utet, 1967).

HAS: Académie Royale des Sciences, Paris. *Histoire* (annual reports).

MAS: Académie Royale des Sciences, Paris. *Mémoires* (collections of papers).

Introduction

THIS BOOK deals with an exemplary episode in the history of science: the controversy between the supporters of Luigi Galvani's theory of animal electricity and its opponents, led by Alessandro Volta. The episode's importance was aptly described by one of the scientists extensively involved in its early aftermath: "The storm aroused by the publication of [Galvani's] *Commentarius* among physicists, physiologists, and physicians can be compared only to the storm that at the same time [1791] arose on the political horizon of Europe" (du Bois-Reymond 1848, 1:49; see Hoff 1936, 159). And the enduring relevance of the controversy is demonstrated by the fact that one of its key aspects is still open to debate.

At the time our story unfolded, scientific writing was expected to offer instruction through amusement. The culture of the Enlightenment fostered the belief that proof and experiments would be more effective if conducted in an entertaining manner, capturing the attention of a nonspecialist public. The dividing line between laboratories and *salons* was not as rigid as today, and the efforts of many eighteenth-century practitioners to make science accessible to the educated public still deserve our admiration. Typical of such efforts was the inclusion of stories, anecdotes, and "tales of philosophy and romance" in manuals and treatises, especially in the introductions; so was the fact that entire books of a strictly scientific nature were addressed to "ladies" or "fräuleins."

To introduce the reader to my subject, I too shall follow this practice. A brief historico-philosophical tale will make it easier to highlight the main events that will be analyzed in depth later on. This will allow me to anticipate the moral of the story. I am generally against drawing moral lessons from history—among other reasons, because history provides no such lessons except those that are deliberately concealed in it (see Pera 1986–87). However, I believe a historical episode can be used to *illustrate* a moral. The latter is ultimately the theoretical viewpoint from which the episode was reconstructed and without which it would not acquire historical status. Naturally, it is the reader who must decide if that viewpoint is suitable to the episode and applicable to others. Philosophers are ever eager to generalize, and historians are right to remind them of the particular.

But here is the tale.

HOW A DISSECTED FROG, WHEN REQUESTED
TO REVEAL ITS INHERENT ELECTRICITY,
MADE A MOCKERY OF A DOCTOR AND
A PHYSICIST, AND REVEALED THEIR
HIDDEN METAPHYSICS INSTEAD

Once upon a time there was a frog. It fell into the hands of a doctor, was sacrificed in the name of science, and "prepared in the usual manner," as that doctor and others with him were used to doing with animals of its species. Those who have forgotten what that manner was can look up the doctor's writings. The hapless frogs were "cut transversally below their upper limbs, skinned and disemboweled . . . only their lower limbs were left joined together, containing just their long crural nerves. These were either left loose and free, or attached to the spinal cord, which was either left intact in its vertebral canal or carefully extracted from it and partly or wholly separated" (GM, 5; GOS, 125).

Thus prepared, the animal lived out its wretched existence (or rather, its final agony) being shuttled from one laboratory table to another. Everyone would look at it, touch it, prick it with pins, needles, and knives, and even torture it with terrible electrical discharges from a recently invented machine. This device was the great pride of contemporary science and a source of considerable entertainment for ladies—and for the gentlemen who courted them in the *salons*. One day, however, the dissected frog decided to avenge itself for the affronts and suffering it had endured.

That day, unbeknown to the doctor, the frog positioned itself so that its residual extremities—nerve and muscle—touched a metal. This caused it to feel a shock all down its body, but as the creature was unable to sigh—despite its undoubted intention of doing so—it drew the doctor's attention by starting to contract. Noticing this, the doctor nearly exclaimed it was a miracle, for no dissected frog had ever been observed to move all by itself, exactly in the same manner as it contracted when tortured with direct discharges of electricity. The physician began to study the phenomenon. In fact, he studied it at length, testing and retesting as prescribed by the rules of the method he slavishly obeyed. Finally, one day, he had an illumination.

"Eureka, eureka!" he shouted. "I have understood the mystery. The frog is endowed with an electrical fluid that is in a state of disequilibrium, condensed in the nerve and rarefied in the muscle. This fluid, conveyed under the law of equilibrium from one region to the other, streams through the nerves, causing the muscles to contract sharply and the leg to rise. At last, I've discovered animal electricity!"

The frog kept up the game, which, after all, had only just begun. To prolong the fun, it placed itself in the hands of a physicist who, like all physicists, stuck to positivity and did not believe in miracles—not even scientific ones. This gentleman liked to repeat the famous phrases that, rightly or wrongly, have ensured the success of his profession throughout the centuries. "In science," he would hold forth, for example, "one must not subscribe to vague principles on the strength of their flattering appearances" (VO, 7:64). One must not multiply entities without justification (VO, 5:82). One must not accept conjectures that do not tally with experiment (VO, 3:6, and VO, 6:274), and one must be ready to give up even the most attractive ideas when experiment refutes them (VO, 6:78). And also the favorite maxim: "The language of experiment is more authoritative than any reasoning: facts can destroy our ratiocination—not vice versa" (VO, 7:292).

The physicist cast a suspicious eye on the frog and began to repeat the doctor's experiments. The frog deliberately made few or no contractions when the physicist touched its nerve and muscle together with an arc composed of a single metal, but jerked "vigorously" when the arc was bimetallic. Confronted with these phenomena, the physicist, whose feeling about the doctor's explanation had already shifted from incredulity to doubt, cast away all remaining suspicions.

"Eureka, eureka!" he shouted, joyfully discarding a useless entity. "I've understood it all. The secret lies entirely in the metals. When metals of different kinds are brought in contact, they cease to be simply good conductors of the fluid. They also act as motors for it. A disequilibrium is established between them: one metal loses the fluid and becomes negative; the other acquires the fluid and becomes positively charged. If the two metals come in contact with a humid body, such as the body of the dissected frog, the fluid circulates in compliance with the law of equilibrium—hence the contractions. These, therefore, have nothing to do with animal electricity; they are due to plain physical electricity!"

"Oh yes?" replied the frog with a defiant gesture. "How do you square that with this phenomenon?" And with these words it returned to the doctor's table and began exhibiting sharp contractions even when the doctor, *without* using metals, merely touched its muscle with its nerve.

"Eureka, eureka!" shouted the doctor, jumping up in triumph. "This is a crucial proof. If mere nerve-muscle contact provokes the contractions, that means the electrical fluid in the frog is naturally unbalanced. So I'm the one who's right: animal electricity exists!"

But the dissected frog had now decided to play the game all the way. It made itself scarce and allowed the physicist to obtain an electrical current by simply piling up a series of suitably combined moist metallic bodies.

"Eureka, eureka!" the physicist thundered again. "Didn't I say so, that animal electricity had nothing to do with it? Here's the crucial proof: an electrical frog that displays all the signs of the other electricity, and in an even livelier manner. In fact, it generates a perpetual motion that has nothing animal about it. What more do you want?"

Indeed, no one seemed to want anything more. The physicist convinced nearly everyone. While the doctor died in poverty, his rival reaped honors, money, prizes, advancement, and prestige. For lack of a Nobel prize (still to be invented), he contented himself with a handsome medal pinned on his chest by no less than Napoleon Bonaparte.

But he failed to convince the frog. "Who told you," the frog insinuated, "that your electrical frog is identical to me—a frog of flesh and blood, that croaks, moves, makes love, is engendered by frogs and engenders other frogs? True, you put two moist bodies together and get electrical signs. But who told you my nerve and my muscle are two *generic moist bodies*? No sir! My nerve and my muscle are *specific organic bodies*, and if I produce electrical signs when I touch them together, my electricity is all mine—distinctive and innate."

The physicist confidently retorted: "What on earth are you saying? Can't you see, for example, that my pile of conductors is analogous to the pile of membranes that the torpedo fish uses for the same purpose, namely, to produce electricity and deliver shocks?"

"Analogy?!" replied the frog. "You're the one who's talking about analogy? But wasn't your favorite slogan 'facts and ratiocination'? Now, the facts are what they are. You see them the way the doctor sees them. And as for ratiocinations, which one is correct? That of the doctor, who, looking at my nerve and muscle, sees just that—a nerve and muscle—and therefore speaks of animal electricity? Or your ratiocination, when, observing the same entities, you see only two common moist bodies and therefore speak of common electricity?"

The dissected frog had not yet vented all its wrath. In its thirst for total vengeance, it turned to the doctor and the physician alike, and carried on:

"Don't you see I'm ambiguous? Don't you understand that the whole secret lies in the way you look at me? And the way you look at me depends neither on facts nor on ratiocination. I'll tell you what it depends on: your hidden metaphysics. I know, you've never thought about it. You think metaphysics is no business of yours—but you're wrong. You keep it around the house without realizing, and metaphysics *is* your business—very much so! Look at me. In the name of one metaphysics, I am one thing; in the name of another metaphysics, I'm something else. Yet I am something, but your crucial experiments will never tell you what. Isn't this splendid? You've used me to reveal my electricity and I'm revealing your hidden metaphysics to you! Three cheers for frogs and down with your art, which tries to demonstrate everything with facts and ratiocination!"

The fable demonstrates that even dissected frogs have a throbbing, philosophical heart. This book tries to find out more than what the fable tells us—even though, like all fables, this one teaches us the way of the world. At any rate, we shall attempt to find out the true course of the events that pitted the doctor and physiologist from Bologna, Luigi Galvani, against the physicist from Como, Alessandro Volta. And we shall also try to understand the reasons, both explicit and hidden, for the dispute.

From the logical point of view, the issue seems simple. Between the statements that "there exists an animal electricity" and "there is no animal electricity," there is no middle course. Either Galvani or Volta is right. Even from the methodologist's point of view, the issue seems simple. For if we add the experiments and proofs ("facts and ratiocination"), we ought to know who was actually right. Yet everything is not so clear.

In reality, the true alternatives are: "There exists an animal electricity" and "there exists a contact electricity between dissimilar conductors." As a result, there is not only a middle course, but even a fourth possibility and more: either Galvani is right, or Volta is right, or both, or there is another solution. This raises an initial problem. If, by logic, a clash between empirical theories never involves taking or leaving one of only two solutions, in practice that is exactly what happens. Why? What is it that drives the situation to a rigid stalemate, "freezing" the antagonists and reducing them to two mutually impenetrable camps? What are the criteria—of merit, of method, or of another nature—that determine the formation of camps and dividing lines in the scientific community? And what ultimately settles the controversy?

Personal factors naturally play a part, such as the emotional commitment and intellectual passion of each scientist with regard to his own discovery. Social factors too are at work, such as the status and prestige of an academic institution or school of thought. In the Galvani-Volta debate, it is hard to deny the influence of either type of factor. We need only look at the protagonists and what they embodied. First, Volta—a highly renowned scientist, a man of many interests, a prolific writer, an esteemed international authority whose many inventions of instruments had won him a large following, a recipient of prizes from the Royal Society and from Napoleon. Against him, Galvani—a cautious, strict-minded physiologist, anatomist and doctor, a shy personality, slow to communicate his findings, a poor correspondent, averse to travel and worldly fame, fervent in his Catholicism but also in his political beliefs, to the point of dying in poverty for having refused an oath of allegiance to the French-imposed Cisalpine Republic. To put it differently: one camp was occupied by a discipline—physics—that was already queen of all the sciences and was expanding daily; in the other camp stood a series of disciplines that were waiting to take off—like physiology—or remained strongly empirical and pragmatic, if not spurious and suspect, such as electrical medicine, not to mention medicine itself. In short, the contendants were grossly unmatched.

And yet, even by drawing on the full resources of the psychology of invention and the sociology of research, we could not exhaust the phenomenological complexity of scientific controversies. Indeed, in this dispute, we would not even grasp the essentials. To begin with, it was a difference in strategy that set frogs against the "magic power of metals." Galvani's strategy was particularly clever. From his very first experiments, he sought to support his conclusions with the known, accepted laws of electricity. He did this not only in the so-called first experiment on contractions at a distance but, more importantly, in the "second experiment," where he observed that contractions could be produced on a dissected frog simply by joining the nerve and muscle together with a metallic arc. As we shall see, Galvani's main argument ran as follows: Let us assume physics is correct in viewing metals as conductors and "balancers" of electric fluid. If so, in the presence of muscular contractions entirely identical to those obtainable by discharges of artificial electricity, physicists should agree that (a) such contractions too are due to a flow of electrical fluid; (b) this flow is caused by a disequilibrium; and (c) this disequilibrium is natural and inherent in the parts of the animal brought into contact with the arc. In substance, by invoking the authority of physics over physiology, Galvani intended to show that a particular physiological apparatus, muscular fiber, could only be equated with a single electrological apparatus—the Leyden jar—*and with no other.* It would then be possible to counter the traditional objections that had been raised, precisely on physiological grounds, against the animal-electricity hypothesis—in particular, the argument that such electricity would be dissipated rather than unbalanced in conducting bodies. The rebuttal would rely on special hypotheses about the distinctive composition of nerve and muscle.

But Volta's strategy was just as able and effective. If Galvani relied on physics, Volta tried to score by throwing him off balance. Volta's attempt consisted entirely of overturning the premise of the argument—*if* physics is correct . . .—from which Galvani derived the theory of animal electricity. To this end, Volta sought to show that physics itself needed revisions or at least supplements. In particular, he tried to show that metals acted not only as conductors of electrical fluid but as motors and "unbalancers." Therefore the disequilibrium that generated muscular contractions in the frog could be explained by a wholly physical law as an effect of the metals in the arc. At this point, an obvious application of Occam's razor should have led to the abandonment of the animal-electricity principle as superfluous, and the frog should have been likened to *nothing other than* an electrometer, albeit the most sensitive one known.

In essence, while Galvani proposed a *naturally unbalanced animal electricity*, Volta countered with *an artificially unbalanced electricity in the animal.* The frog was ambiguous and lent itself to both interpretations. It is, however, a rule of methodology—sometimes followed by scientists—that a theoretical interpretation of empirical experiments remains little more than a plausible hypothesis so long as it is supported by those experiments alone. Nor does the situation change much if the theoretical interpretation proves compatible with other accepted laws and theories. Being aware of this, both Galvani and Volta had to seek other proofs, preferably crucial ones. As we shall see, the two men found them, and the controversy developed from a conflict of strategies into a conflict of experiments as well.

By simply touching the frog's nerve and muscle together *without* metals—the so-called third experiment—Galvani produced contractions. This was a crucial experiment. Volta admitted as much, but did not give up. First, he transformed the theory we shall refer to as the *special* theory of contact electricity—*metals* are the only electromotive substances—into a *general* theory of contact electricity, which stated that all *conductors*, metallic or not, are electromotive. He then proved that pairs of suitably combined metals or conductors generated an electrical current—indeed, in the case of the pile, a "perpetual" current. And this too was a crucial experiment.

As a result, the frog became more ambivalent than ambiguous. How was one to accept this impossible situation? Two crucial experiments, one apiece for two incompatible theories, is more than logic and methodology can admit. Yet both experiments were unequivocal. The situation could therefore indicate two alternatives: either both theories—both sides of the frog—were true, complementary, and only apparently at odds; or their opposition was real but the two theories were observationally equivalent, like a single frog from two different perspectives.

It is no help to us to know that the first alternative is the one we now regard as correct; to know that a bimetallic contact generates the so-called Volta effect (see the writings of Luigi Giulotto collected in Giulotto 1987 and the bibliographical references there), and that there exists an "injury current" of the nerve or muscle; or to know that contact between metals and animal tissues produces an electrochemical reaction (this is the third alternative, which, although already formulated at the time, was never taken into consideration by the two adversaries). The theories thus reconciled today are no longer those that were then in conflict. True, at a certain point in the controversy, the hypothesis of a compromise based on an even distribution of explanatory principles was advanced by both sides. The circumstances and the very mode in which the compromise was refused show us, however, that the two positions were truly incompatible and informed by an "all or nothing" logic. As Galvani wrote: "He, in short, attributes everything to metals, nothing to the animal; I, everything to the latter, nothing to the former, as far as the imbalance alone is concerned" (GO, 303–4; GOS, 429).

The second alternative remains. But what does it mean to say that the theory of animal electricity and the general contact theory were observationally equivalent? It means that they had the same empirical consequences; that there were no truly crucial proofs to confirm or invalidate either theory—despite appearances suggesting even two such proofs; lastly, that, for lack of such proofs, the acceptance or rejection of the theories was dictated by extra-empirical, if not necessarily extra-rational, factors.

This leads us to another dimension of the controversy. More than a conflict of strategies or experiments, it was a clash of assumptions or interpretative theories that functioned as gestalten. In particular, it was a conflict between an electrobiological gestalt and an electrophysical gestalt in the same domain of observation. In the course of our discussion, we shall try to show how these gestalten predetermined the domains to which phenomena were assigned, how they selected the different kinds of explanatory hypotheses, how they each suggested the adjustments of these hypotheses to their respective anomalies, and finally how—other things being equal, that is, when all the proofs were exhausted—they single-handedly determined which of the two contending hypotheses would be accepted.

When the conflict had reached this stage, Volta was able to transform defeat into victory. His special theory of contact electricity—the metallic-electricity theory—had been confuted by Galvani's experiments and rejected. But his general theory of contact electricity could not be similarly invalidated, for the simple reason that it stated the same things as the animal-electricity theory. It did so, however, not with *different words* (as Galvani sensed and Volta, instead, failed to see), but rather with *different concepts*, with a different interpretation about the realm to which the phenomena belonged—in short, with a different theoretical gestalt. It was on this ground of hidden metaphysics rather than that of proof and logic ("facts and ratiocination") that the frog's ambiguity disappeared in Volta's favor. Thus, if we are telling the truth, so is the fable.

THE AMBIGUOUS FROG

Who can fail to see that we're
both saying the same thing, albeit
with different words?
(Luigi Galvani, *Opere edite e inedite*,
1841)

CHAPTER 1

Electricity, the Science of Wonders

1.1 THE EIGHTEENTH-CENTURY VOGUE

Who knows what particular examples old Aristotle had in mind when he said that science begins with wonder. Electricity was almost certainly not one of them. Not that the philosopher was unacquainted with the properties of amber—or, as it was called in his language, ἤλεκτρον. By the time of Thales of Miletus (sixth century B.C.)—and possibly earlier—it was known that this stone, if rubbed even just with a dry hand, behaved oddly like a magnet, attracting bits of straw, dry leaves, and other light bodies. But, although familiar, these properties had never been methodically investigated. Nor had they ever prompted the emergence of an activity even remotely resembling a science, as occurred more than two thousand years later.

If we may be allowed a small historical compression, we can say that the story of electricity truly began around the early eighteenth century, around the 1720s to be exact. It is not hard to understand why electricity should have chosen precisely the Age of Enlightenment to reveal its wonders. Broadly speaking, this was due to two sets of circumstances—a recurrent phenomenon in the history of science. First, new curiosities and experimental data; second, an intellectual infrastructure capable, if not of mastering them, at least of providing a plausible theoretical framework. This proves the point that no fact, however strange and curious, becomes relevant except in the context of a given problem, and no problem is raised or attracts attention unless it exhibits theoretical implications.

The fact remains that around or shortly after 1720 electricity established itself as a topic of dominant interest, chiefly thanks to the work of Stephen Gray, then a pensioner of the Charterhouse in London, and the research undertaken shortly thereafter by Charles Du Fay, a polymath scientist and keeper of the royal gardens in Paris. From that moment on, all of cultured Europe was electrified—not only in the metaphoric sense—and astounded by the steady flow of new wonders. Indeed the initial, enduring, unanimous reaction was sheer wonderment. Wherever electricity was talked and written about—in books, memoirs, articles, letters, journals; in other words, all over—it was described as "wonderful." Even a cursory glance at the literature reveals the proliferation of this sentiment.

To begin with, it was the effects that were wonderful. The "wonderful effects" of electricity, wrote Carlo Taglini (1747, 153n); the "wonderful effects of electricity," exclaimed Giuseppe Veratti (1748, preface), the Abbé Jean-Antoine Nollet (1749b, 446), and Tiberio Cavallo (1781, 1) in chorus. Each novelty was greeted with the same refrain: the "wonderful effects of pointed bodies" (Benjamin Franklin 1941, 171); the "wonderful effects" of the electrical battery (Cavallo 1782, 58); the "wonderful effects of animal electricity" (VO, 1:15); the "wonderful effects of contractions" (Giovacchino Carradori 1793, letter 4:3); the "wonderful effects of the pile" (Jean-François-Dominique Arago 1854, 222n; Antonio Cima 1846, 3).

Next came electrical phenomena, facts, and "virtues." Robert Symmer investigated not electricity but its "wonderful operations" (1759, 371)—as did everyone else. Nollet described the "wonderful electrical phenomena" (1746a, 21) or the "wonderful phenomena of electricity" (1748a, 164). Symmer also wrote that "what appears wonderful to us, is, that bodies should be at all capable of acting upon one another at a distance" (1759, 388). Reporting his first famous experiment, Galvani spoke of the "constant and wonderful phenomenon" of the frog's remote contraction (GM, 254); concerning another experiment, he noted the "wonderful fact" of contractions with a bimetallic arc (GM, 62), while the animal's muscular motion struck him as a "truly wonderful movement" (GM, 64). In spite of his utter disagreement with Galvani, Volta fully shared the latter's sense of wonderment. For example, in refuting his opponent's experiments, Volta cited the conductors' "wonderful property of stimulating the electrical fluid" (VO, 1:297), which corresponded to the "wonderful property" of electric fish "of imparting shock" (VO, 2:296)—a far cry from the "amusing and wonderful little games" (VO, 4:470) that one could play with artificial electricity.

There was also a host of discoveries and experiments. The "wonderful experiment" of drawing down lighting, said Veratti (n.d.); the "most wonderful electrical experiments," asserted Doctor Giovanni Fortunato Bianchini (1749, 15); the "wonderful cures," proclaimed Joseph-Aignan Sigaud de la Fond (1785, 473); the "wonderful technique" of purging by electricity, wrote Veratti (1748, 123); the two states of electricity "so wonderfully . . . combined and balanced," exclaimed Benjamin Franklin (1941, 181); the "wonderful movements of bodies immersed in a different atmosphere," said the editors of Giambattista Beccaria (1793, xii); the "wonderful and important discovery" of the electrical effects of metallic pairs, announced John Robison (1793, 174); while Volta—at the time he believed or pretended to believe in it—referred to the "wonderful discovery of animal electricity innate in and specific to organs" (VO, 1:47n) and Emil du Bois-Reymond to the "wonderful revival of severed limbs" (1848, 1:51).

Instruments were no exception either. The Leyden jar, the first condenser in the history of electricity, became "M. Mus[s]chenbroek's wonderful bottle" (Franklin 1941, 179); the electrophore was a "most wonderful invention" (VO, 3:99), a "wonderful discovery" (Fromond in VO, 3:112n); the pile nothing less than "the most wonderful instrument ever created by the human intellect" (Arago 1854, 228) or "the most wonderful instrument ever invented by mankind" (pp. 219–20); or, simply, the "wonderful instrument" (André-Marie Ampère 1958, 100).

Not even laws and theories escaped this vortex of wonders. Ulrich Theodor Aepinus wrote of the "wonderful phenomena and laws of electricity" (1759, 1). Volta spoke of Franklin's theory as the "wonderful theory we are following" (VO, 4:359) and referred to his own theory as "wonderfully" confirmed (VO, 3:155).

In short, everything about electricity was wonderful. The "wonders of electricity," exclaimed Nollet (1746b, 129)—echoed by Franklin (1945, 750), Gianfrancesco Pivati (1716, 3:500), Francesco Griselini (1748, 41), and Guyot (1786, 217). And since electricity itself was "replete with wonders" (Jean-Paul Marat 1782, 355), "a subject in which wonders are continually encountered" (Du Fay 1733a), in which "we can expect new prodigies daily" (Jean Jallabert 1748, v), we can understand how it soon turned from a "small phenomenon in physics" (HAS, 1733, 4) into the "favourite object and pursuit of the age" (Cavallo 1782, 1), the "fashionable topic" (Marc'Antonio Caldani, in Caldani and Lazzaro Spallanzani 1982, 280). In the words of Joseph Priestley, "electricity has one considerable advantage over most other branches of science, as it both furnishes matter of speculation for philosophers, and of entertainment for all persons promiscuously" (1775b, 2:134).

Indeed there were attractions of every kind: bodies that repelled rather than attracted each other; cold metals and even water and ice that gave off sparks, violent and potentially lethal shocks, flashes, thunderlike explosions, and even harmless lightning. In other words, "Here we see the course of nature, to all appearance, intirely [sic] reversed" (ibid., 135). One might well ask,

> What would the ancient philosophers, what would Newton himself have said, to see the present race of electricians imitating in miniature all the known effects of that tremendous power [of thunder and lightning], nay disarming the thunder of its power of doing mischief, and without any apprehension of danger to themselves, drawing lightning from the clouds into a private room, and amusing themselves at their leisure, by performing with it all the experiments that are exhibited by electrical machines. (pp. 135–36).

Newton, Priestley implied, would have been astonished. Why not, then, his successors and—even more so—the masses, or people with only a modest education? In fact, that is precisely what they did. Within a short time the fashion exploded. No sooner had they discovered the new wonders than science buffs, dilettantes, practitioners, ladies, gentlemen, fräuleins,[1] officers, clergymen, poets, and even quacks and charlatans,[2] in short, everyone from intellectuals to their victims vied to illustrate or experiment with the pranks—indeed "the wonderful pranks"—of electricity (Sguario 1746, 28). Nor did scientists stay away from the game. On the contrary, they were the first to rush from the austere lecture halls of academe and the arid pages of journals into *salons* and gardens.

So seriously did scientists take this popularization and entertainment, and so far were they from dismissing it as just a society game, that they themselves made a principle of their duty to instruct through amusement. This was clearly stated by Eusebio Sguario, who, having fully digested the lesson of his compatriot Francesco Algarotti and of Voltaire, published an anonymous work that we may describe as "Electricism for Ladies"—although, in its scientific rigor and theoretical originality, it is anything but contemptible. "In our century," he wrote in the preface to his *Novella filosofica e galante, che serve d'introduzione alla dottrina delle forze elettriche,* "it is indeed a crime for a Writer to deal with boring topics that tire the reader into yawning; it is desirable that the yet more abstract, elevated, and sublime sciences should be endowed through the writer's ingenious craft with some pleasing adornments and enjoyable, lighthearted treatises" (Sguario 1746, vi–vii).

Forty years later, Guyot, a member of the Société Littéraire et Militaire de Besançon, voiced the same notion, in the hope that his *Nouvelles récréations physiques et mathématiques, contenant ce qui a été de plus curieux dans le genre, et ce qui se découvre journellement* could "steadily augment the enlightened century's taste for the study of experimental physics." Guyot added:

> If that study is so often full of thorns, an effort to cover it with flowers is perhaps another way to promote its cultivation. Some authors have succeeded in making certain works on physics both attractive and interesting by using an easily understandable method within everyone's reach. These works have always been greeted with great favor: *omne tulit punctum qui miscuit utile dulci* [he who combines utility and delight scores every point]. (1786, 219)

Stephen Gray and Charles Du Fay were the first to espouse this belief. Not only did they discover properties and laws of electricity, but they also contributed to recreational popularization. Gray's famous experiment of 1730 consisted in suspending a child from ropes and electrifying him. Regularly repeated across the ocean by Franklin and his collaborators at the Library Company in Philadelphia, and across the Channel by Du Fay at the court of Louis XV, it never failed to offer a surprising spectacle. Using a rubbed glass tube, Gray had begun by electrifying everything from "a Guinea, a Shilling, a Half-Penny, a Piece of Block-Tin, a Piece of Lead" (1731, 21–22) to "a Fire-Shovel, Tongs, and Iron Poker, a Copper Tea-Kettle . . . whether empty, or full of either cold or hot

1.1 Gray's experiment. From Nollet 1746b.

water, . . . Flint-Stone, Sand-Stone, Load-Stone, Bricks, Tiles, Chalk, . . . several vegetable Substances, as well green as dry" (p. 22), and later "a large Map of the World," "a Table-Cloth," "an Umbrello," and "a live Chick" (pp. 31–33). One day, on 8 April 1730, he came up with the idea of trying the experiment on a human body. He took an unlucky forty-seven-pound boy, suspended him horizontally from the ceiling with ropes, electrified him by bringing the glass tube up to various parts of his body, and observed that the boy attracted leaf-brass (see figs. 1.1 and 1.2).

When Du Fay repeated the experiment, he made another discovery. Suspending himself "on the Lines," he observed that

if the electrick Tube be put near one of my Hands, or my Legs, and then if another Person approach me, and pass his Hand within an Inch or thereabouts of my Face, Legs, Hand or Cloaths, there immediately issues from my Body one or more pricking Shoots, with a crackling Noise, that causes to that Person as well as to my self, a little Pain resembling that from the sudden Prick of a Pin, or the burning from a Spark of Fire, which is sensibly felt thro' ones Cloaths, as on the (bare) Hand or

1.2 Gray's experiment and the ignition of alcohol. From Sguario 1746.

Face. And in the Dark these Snappings are, as may be easily imagined, so many Sparks of Fire. (1734, 261–62)

At this point, the road was open and the race was on to develop the most singular experiments. In 1743, Georg Mathias Bose, professor of theology at Wittenberg, performed the "beatification," so called because

[he] placed a very young boy on the resin cakes; or, when the boy was biggish, in a fairly large box with tall sides, smeared on the inside with enough resin to cover the boy halfway up his legs; [Bose] then electrified him at length with the globe, on which the boy placed one of his hands. Before long, the electrical matter accumulated in such immense quantities inside him that first his shoes, then his legs and knees appeared to be covered in fire. Finally his entire body was bathed in light and surrounded in the manner sometimes used to depict the glory of a saint by encircling him in rays of light. (Sguario 1746, 268–69)

Two years later, Bose switched from "beatification" to "coronation": "A Chair being suspended by Ropes of Silk, made perfectly dry, a Man placed therein is render'd so much electrical by the Motion of the above-mentioned Globe, that, in the dark, a continual Radiance, or *Corona* of Light, appears encircling his Head, in the manner Saints are painted" (1745, 420).

Then there was the electrical table setting, which consisted in electrifying a table and watching with amusement as the guests leaped and shrieked when the sparks flew from the fork tips; and the "electrified Venus," a lady—presumably a "fräulein"—who gave stinging electrical kisses. Bose became a virtuoso of this kind of frivolity, and there was heavy demand for his performances all over Europe. In addition to his experimental skills, he was also, unquestionably, a consummate entertainer. At one of the many scientific-social occasions in which he took part, he declared:

> Gentlemen and dear friends, you who grace electricity with your presence, you have before you a wide array of temperaments, on which you will be able to observe the great variety of electricity's actions. Rid your mind of all unhappy thoughts: this place is the enemy of melancholy. Hypochondriac vapors, boredom and sadness are strictly forbidden in this hall. Here each is free to make the comments he most pleases, in jest and mirth. Pay attention, and don't disturb the show. (Quoted in Sguario 1746, 44)

Where Bose failed, his imitators succeeded, such as Christian Friedrich Ludolff of Berlin, who used an electrical spark to ignite alcohol in a spoon (fig. 1.2).

Apart from a rapid advancement of knowledge in the form of laws and theories, which we shall examine later, one of the first effects of these experiments was to promote the introduction and improvement of electrical machines (see fig. 1.3). Within some sixty years, these graduated from Francis Hauksbee's rudimentary glass tube,[3] still used by Gray and Du Fay, to the more complex and efficient glass-disk machines of Jan Ingenhousz and Jesse Ramsden, also employed by Galvani and Volta.

A second effect has a deeper theoretical and practical impact: the invention of the first condenser, the Leyden jar (fig. 1.4a, b).[4] One day, in late 1745, Pieter van Musschenbroek—"naivety itself," in Voltaire's words (see Heilbron 1979, 313n)—was performing experiments in his laboratory at Leyden in the company of Andreas Cunaeus and Jean-Nicolas-Sébastien Allamand. The three were attempting to fill a bottle with electrical fluid gathered by means of a metal wire from the prime conductor of an electrical machine. At one point, Cunaeus, who was holding the bottle in one hand, touched the conductor with his other hand and felt a powerful shock. The same phenomenon had occurred, almost in the same period but unbeknown to the Leyden trio, to Ewald Jürgen von Kleist in his Leipzig laboratory. Kleist was seeking to augment the sparks by increasing the size of the conductor from which they were drawn. He was expecting to achieve this by introducing into a small bottle of water a metal wire connected to the machine's prime conductor. But here is a description of the two experiments.

(a)

1.3 The development of electrical machines: (a) Guericke's machine (from Guericke 1672); (b) a variant of Hauksbee's machine (from Sguario 1746); (c) another variant (from Griselini 1748); (d) Bose's machine (from Nollet 1746b); (e) a variant with bundles of various conducting materials between the globes and the prime conductor and with Winkler's cushion (from Priestley 1775b); (f) cylinder machine (from Sigaud 1785); (g) Ramsden's machine (from Guyot 1786).

Kleist's is more measured and technical:

When a nail, or a piece of thick brass wire, &c. is put into a small apothecary's phial and electrified, remarkable effects follow: but the phial must be very dry, or warm. I commonly rub it over before-hand with a finger, on which I put some pounded chalk. If a little mercury or a few drops of spirit of wine, be put into it, the experiment succeeds the better. As soon as this phial and nail are removed from the electrifying glass, or the prime conductor, to which it hath been exposed, is taken away, it throws out a pencil of flame so long, that, with this burning machine in my hand, I have taken above sixty steps, in walking about my room. When it is electrified strongly, I can take it into another room, and there fire spirits of wine with it. If while it is electrifying, I put my finger, or a piece of gold, which I hold in my hand, to the nail, I receive a shock which stuns my arms and shoulders. (Quoted in Priestley 1775b, 1:103)

More emotional and geared to dramatic effect was Musschenbroek's account:

I would like to tell you about a new but terrible experiment, which I advise you never to try yourself . . . I was engaged in displaying the powers of electricity. For this purpose, I had suspended an iron tube AB from two blue-silk lines, fig. 1. A glass globe was rapidly spun, rubbed by hand, and its electricity communicated to the tube. At the other end B a brass wire hung loosely. Its tip was dipped into a

(b)

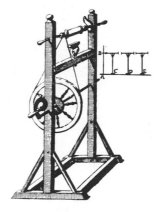

(c)

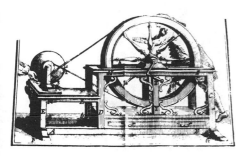

(d)

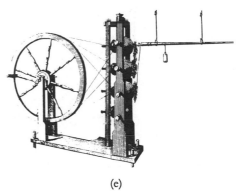

(e)

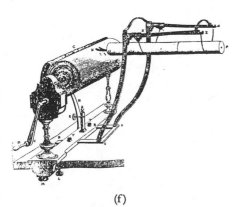

(f)

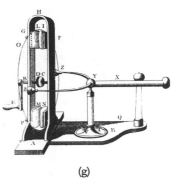

(g)

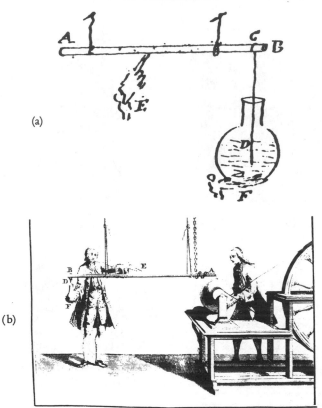

1.4 The Leyden jar: (a) Nollet's ms. sketch, from Benguigui 1984;
(b) from Nollet 1746a.

round glass jar D, partly filled with water, which I held in my right hand F; with my left hand E I tried to draw sparks from the electrified iron tube. Suddenly my right hand F was struck with such force that my body shook as if hit by lightning. Generally the blow does not break the glass, no matter how thin it is, nor does it knock the hand away; but the arm and the entire body are affected so terribly I can't describe it. I thought I was done for. (Quoted in Nollet 1746a, 2; trans. adapted from Heilbron 1979, 313–14, who uses the manuscript proceedings of the Académie des Sciences)

The experiment was indeed curious and terrifying. All who repeated it—and they were many, as soon as the news spread[5]—were both amazed and worried. Allamand wrote to Nollet: "The first time I tried it, the blow left me breathless for a few moments. Two days later, Musschenbroek tried it with a glass phial and was so stunned that when he came to see us a few hours afterward he was still shaken and told me that nothing in the world would persuade him to repeat it" (quoted in Nollet 1746a, 3).

Johann Heinrich Winkler wrote:

> When I heard of Mr. Musschenbroek's Experiment, I tried the same; but I found great Convulsions by it in my Body. It put my Blood into great Agitation; so that I was afraid of an ardent Fever; and was obliged to use refrigerating Medicines. I felt a Heaviness in my Head, as if I had a Stone lying upon it. It gave me twice a Bleeding at my Nose, to which I am not inclined. My Wife, who had only received the electrical Flash twice, found herself so weak after it, that she could hardly walk. A Week after, she received only once the electrical Flash; a few Minutes after it she bled at the Nose. (1746, 211–12)

The pranks of electricity, while wonderful, could therefore be dangerous too. "Our better-informed descendants," Nollet wrote, "may laugh at our fears, but if they are fair they cannot blame us for the reason that forces me to voice this warning, however superfluous it may turn out to be" (1746a, 23). Yet, despite the greater caution, the prevailing attitude remained one of amusement. Indeed, the Leyden jar had a multiplier effect on curiosity and parlor tricks. The Abbé Nollet became the acknowledged master of this new kind of experiment. Jean-Antoine Nollet was a pupil of Du Fay, a polymathic scientist, member of the Académie des Sciences and the Royal Society, tutor to the Dauphin, and "experimenter" to Louis XV. One of Europe's leading "electricians"—as students of electricity were then called—Nollet was a true "prince of science" who excelled in the art of experiment and entertainment. One day, he discharged an electrical shock among "200 men, lined up in two rows each more than 150 feet long" (1746b, 135; see also 1746a, 18). On another occasion, in the Grande Galerie at Versailles and in the king's presence, he jolted 180 soldiers holding each other by the hand (Priestley 1775b, 1:125; Torlais 1954, 81). Later still, at the Collège de Navarre, he surpassed himself by setting up a chain of over 600 people, all of whom felt the shock throughout their body with a violence proportional to the distance of each organ from the point of contact (Sigaud 1785, 237). In all these experiments, the first man in the chain held the bottle while the last man touched the prime conductor.

These attractions proliferated. Louis-Guillaume Le Monnier gave an electrical shock to 140 people (fig. 1.5),[6] suggested a more convenient way to charge the jar, and found that the shock was greatest "when the bodies we want to electrify form part of an imaginary curve stretching from any point on the external iron wire to a point of the bottle below the water level" (1746, 449). The fictive curve was then replaced by a true metal-wire circuit up to 2,000 toises (two and a half miles) long (fig. 1.5). Pursuing this line of inquiry, and for the additional purpose of measuring the velocity of electrical fluid, Le Monnier ran his metal wire through the Tuileries basin in Paris, while William Watson, in

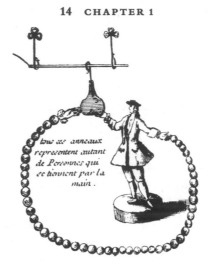

1.5 Experiments with the Leyden jar: the human chain ("all of these rings represent an identical number of people holding one another by the hand"). From Mangin 1752.

London, on 14 and 18 July 1747, conveyed the shock across the Thames, "making use of the water of the river for one part of the chain of communication" (Priestley 1775b, 1:131–37).

During the same period, experiments were also being conducted in the British Colonies. "American Electricity" (Franklin 1941, 176) had arrived a few years late, but soon made gigantic strides ("Who would ever have imagined that Electricity would have learned cultivators in North America?" [Veratti n.d., 1]). Initially, it was pursued by a few dilettantes who performed the usual experiments for money (see Cohen 1941a, 47–56), but very soon it passed into more expert hands. On 28 March 1747, Benjamin Franklin received in Philadelphia a glass tube and a few instructions for its use from Peter Collinson in England. By 11 July, Franklin and his collaborators—Ebenezer Kinnersley, an unemployed Baptist minister but the "principal" and most inventive member of the group (see Cohen 1941b; Lemay 1961), Thomas Hopkinson, and Philip Syng— were able to inform Collinson of the discovery of the "wonderful effects of pointed bodies."

The discoveries were followed by recreations (fig. 1.6). The first was the "counterfeit spider" (Franklin 1941, 177), which danced between the ends of the charged bottle's inside wire and the wire attached to the outside coating "in a highly entertaining manner appearing perfectly alive to persons unacquainted." Next came the electrical book, whose gold line on the cover, after receiving a discharge from the bottle, appeared "a vivid flame, like the sharpest lightning" (p. 186). Then there was the "magical picture," "a large metzotinto with a frame and glass, suppose of the KING (God preserve him)," which, by

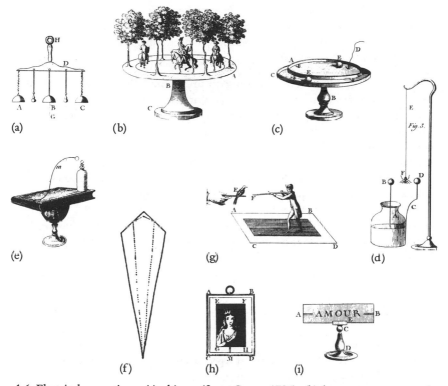

1.6 Electrical recreations: (a) chimes (from Guyot 1786); (b) horse merry-go-round (from Guyot 1786); (c) orrery (from Guyot 1786); (d) spider (from Guyot 1786); (e) book (from Franklin 1773); (f) fish (from Franklin 1773); (g) hunter (from Guyot 1786); (h) crowned king (from Guyot 1786); (i) shining picture (from Guyot 1786).

means of a series of gilt contacts, delivered "a terrible blow" to anyone attempting to remove the crown from the king's head ("If a ring of persons take the shock among them, the experiment is called, The Conspirators") (pp. 193–94). This was followed by electrical spinning wheels, chimes, and orreries (pp. 194–97). Then came the electrical fish, a thin gold leaf tapered to "an acute point": "If you take it by the tail, and hold it at a foot or greater horizontal distance from the prime conductor, it will, when let go, fly to it with a brisk but wavering motion, like that of an eel through the water" (p. 226). To recover from all these labors, the experimenters even planned an all-electric picnic on the banks of the Schuylkill in Philadelphia:

Spirits . . . are to be fired by a spark sent from side to side through the river, without any other conductor than the water, an experiment which we some time since performed, to the amazement of many. A turkey is to be killed for our dinner by the *electrical shock*, and roasted by the *electrical jack*, before a fire kindled by the *electrified bottle*: when the healths of all the famous electricians in *England, Holland, France* and *Germany* are to be drank in *electrified* bumpers, under the discharge of guns from the *electrified battery*. (Pp. 199–200)

While such entertainment was going on in America, Europe was no less active. All the experiments were reproduced or reinvented independently, with variants, simplifications or enhancements. This was the case with the orrery, the merry-go-round, the hunter, and the sparkling picture. The latter served as a device for writing a few favorite words in scintillating letters, usually—and understandably, given the presence of young ladies—"amour."

In addition to such parlor recreations, there were outdoor experiments. Here amusement occasionally combined with utility, yielding interesting discoveries. Perhaps the most dramatic was the discovery made and recounted by Sigaud (1785, 230–33). One day, he was about to conduct a routine experiment. He had formed a chain of sixty people in a courtyard and, after charging the bottle, brought it up to the first in line while the last person drew the spark. Sigaud was expecting the familiar effect. Instead, things went differently. In his words:

> Although the bottle had been very highly charged, the shock was not felt by more than half a dozen people, standing next to the person drawing the spark and to the one holding the bottle opposite. Without changing the arrangement of the chain, I recharged my bottle, but despite the even stronger charge, the effect remained the same and the shock still stopped at the same person, the sixth in the chain on the side of the person drawing the spark.

At this point, Sigaud's bewilderment is understandable. There was clearly something wrong in the chain, and, whatever the problem, it concerned the person at whom the shock stopped: "Everyone took an interest in this person and pretended the incident was due to his particular constitution. The ensuing commotion forced me to abandon the experiment, which I ought to have repeated after removing the person in question from the chain." But what was the matter with that person? It was already known that not all individuals are electrified in the same manner. Sguario, for example, had found that women were electrified better than men, and young people better than old people (1746, 268). As the person involved happened to be a youth, a mischievous hypothesis was inevitable: "After a long time it was assumed the young man in question was not endowed *with everything that constitutes the distinctive character of a man*." But the scientific method requires a hypothesis to be confirmed by rigorous observation before being accepted. And Sigaud followed the scientific method scrupulously:

> Since certain educated people accustomed to experimenting with electricity had assured me some time afterward that it was impossible to electrify or impart shocks to those toward whom nature had been so cruel, I believed I could venture this

observation in one of my courses. However, I stated it not as a fact but a suspicion to be verified. The rumor quickly spread throughout Paris; and while everyone repeated the observation in his own manner, someone assured me that it had been confirmed by a recent experiment performed on a famous musician whom nature has endowed with a bewitching voice and exquisite taste as compensation for the sorry state to which he has been reduced.

Gossip, however, is not scientific proof. A crucial experiment was needed. The request was voiced by the Duc de Chartres, who overcame Sigaud's reluctance. One can understand the apprehension of the experimenter, the spectators and, above all, of the control group:

I was put in charge of the operation and carried out the experiment in February 1772, before many scientists invited by the prince, on three musicians summoned for the purpose. All three felt the effects of the shock and did not in the least impede its travel at any point along the chain. This was composed of about twenty people led by the prince and all felt the shock. The three musicians seemed if anything more sensitive than the others who felt the shock with them; but this greater sensitivity is not baffling. It was certainly due to the surprise of feeling an impulse they had never experienced before, since they had no notion of electricity.

What was to be done? To conclude that the hypothesis was false? There were grounds for that. On the other hand, as scientists knew even without reading Popper, a hypothesis disproved by facts could always be immunized by other, ad hoc hypotheses. While determined to preserve the falsification rule, hapless Sigaud had to take account of this stratagem:

Admittedly, there was no lack of reasons for quashing the rumors that had spread and for showing that the opinion ventured was false; but there were still those who are not easily convinced and do not readily give up an idea they are pleased with and have adopted. They argued that there had to be a difference between men who have been mutilated by Art and men toward whom Nature has displayed cruelty. They consequently held that if the former were able to feel the effects of the shock, the latter might well be unable to do so.

The matter now dragged on and there was no way of silencing the diehards and gossips. Finally "chance," the handmaiden of the scientific method, intervened to clear up the confusion. One day in July, Sigaud was repeating the same experiment with a chain of sixteen people. Once again, the discharge stopped at a certain point. He tried again twice, with no result. This time, however, Sigaud disregarded the constitution, voice, and attributes of the circuit breaker, who was presumably starting to sweat with anxiety. Sigaud looked instead at the poor fellow's legs and noticed that they were stuck in a fairly damp spot. That explained the mystery: the drenched ground was a better electrical conductor than the bodies of the people in the chain. Therefore, when the electricity reached the person who had his feet in the water or just about, it traveled down his legs and was discharged into the ground. Thank goodness. The case was solved and the electrical reputation of the Royal Chapel's musicians—as well as their colleagues throughout the ages—was saved. Electricity is not a reliable test for "white" voices.

1.2 ELECTRICAL MEDICINE: THEORY AND PRACTICE

How long does it take before a physician discovers that a substance, a discovery, or an invention made for entirely different purposes is also good for one's health? Not long, and often—as with electricity or its contemporary, mesmerism—next to no time. (We shall disregard mesmerism here, which ran into immediate difficulties despite the fanatical support it aroused [see Sutton 1981].) There are many witnesses to the fact that the notion of applying electricity to medicine occurred immediately. Indeed, given the prevailing attitudes, it was a reasonable idea. "However interesting the study of physics," an anonymous reviewer of the "state of the question" in electricity wrote in 1746, "it would deserve neither public esteem nor the labors of physicists if it led merely to curious speculations that could contribute nothing to the good of society. It was therefore perfectly natural to conceive the notion of deriving benefits from electricity" (*HAS*, 1746, 8). Poor Plato would have turned in his grave, but the Abbé Nollet shared this view. "When an innovation appears in physics," he wrote in the same year, "curiosity immediately takes hold of it, has fun with it, but is quickly sated. Curiosity then gives way to interest and to the demand that the object of admiration also prove its usefulness" (1746a, 18).

But the most explicit evidence on the matter is perhaps that of Eusebio Sguario, again in 1746:

> Men of science seek not only the pleasant but the useful, which they in fact prefer. No sooner was the great power of electricity over human bodies known than research began with the aim of discovering if it could by any chance alleviate the failings of health. No thought could occur more readily than this, the moment people saw such light flashing from the body, limbs, and skin, and felt the stings, painful blows and sharp stimuli that penetrated almost into the bones when the light appeared. (1746, 366–67)

It is impossible to say who first conceived this noble "thought," but it is not hard to guess where the notion caught on immediately, giving rise to a widespread practice. The movement "started in Italy," although it "made very little progress" there, because, as is well known, "the Italians—who get easily excited about new discoveries and will just as easily fail to investigate them further, mainly for lack of Patrons—no longer thought to repeat the experiment." That, at least, was the opinion of Doctor Giovanni Vivenzio (1784, 6 and 12), "Knight of the royal and military Constantinian order of Saint George, chief physician to Their Majesties," etc., etc.

Let us begin, then, with Italy. The enthusiasm there was such that no fewer than three schools of electrical medicine developed almost simultaneously. On the first school, which employed only the electrification method, we have a detailed description by Giuseppe Veratti of Bologna. In *Osservazioni fisico-mediche intorno alla Elettricitá* (1748), he reported and discussed fifteen cases of cures— authentic "clinical cases"—of which ten were treated with the electrified tube alone. Naturally, the results were prodigious, or almost. The second school, which advocated the method of *intonacature* or "coatings," had as its undisputed master Gianfrancesco Pivati, jurisconsult and physician based in Venice. The author of *Nuovo Dizionario Scientifico* (1746), *Lettera della elettricità medica* (1747), and *Riflessioni fisiche sopra la medicina elettrica* (1749), Pivati sought to "introduce a topical medicine into the innermost parts of the human body" (1747, xxv) by mixing the substance or delivering it via the electrical fluid. His aim was to administer drugs in a manner that preserved them from the alterations they underwent when given orally. This led him to invent the *intonacature*, sealed glass tubes or globes whose inside walls were lined with the medicine. After electrification, the substances would be emitted in the form of fragrant exhalations that would infuse the appropriate part of the body. Here again, at least according to Pivati, the results were excellent.

The method of the third school, almost a synthesis of the two preceding ones, was developed by Giovanbattista Bianchi, "collegiate doctor of medicine," etc., etc., "dean of the medical faculty in the states of H.M. the King of Sardinia" (see Fabri 1759, iii), as well as "known to the republic of letters for his prolific and learned output" (Fabri 1757a, viii). One day—in a truly vivid example of the "theory-ladenness" of observation—"he noted that when persons undergoing electrification were made to hold purgatives in their hands, the extremely subtle particles of those substances were introduced into our bodies, where they produced the effects they usually do when taken orally" (Veratti 1748, 107–8). From the examination of such a singular case to the notion that a medicine held by a person during electrification could enter the body and produce therapeutic effects, the induction was easy if not brief. One of those to fall for it was Veratti, who naturally concluded that the method offered a "new manner, no less wonderful than convenient, to purge when needed those patients who have trouble tolerating the action of orally administered purges" (p. 124).

If we believe the evidence, benefits and healings of this kind were proliferating. Electricity cured not only constipation, but also sciatica, headaches, rheumatisms, herpes, lacrimation, and nervous disorders. Few ailments resisted the therapy. The impact must have been considerable: "All the Italian papers," Sigaud disapprovingly reported (1785, 473), "were filled with a host of miraculous healings that had no other basis than the enthusiasm, not to say the bad faith, of the people who had them published." Outside Italy, of course, the news spread fast. In France, Nollet perked up his ears at once. The first to apply the Leyden-jar shock to the treatment of paralysis, he had diligently applied himself to conducting accurate experiments.[7] In 1748, he experimented with animals before moving on to people. At the Hôtel Royal des Invalides, he secured three paralytics whom he treated with the three most widely used methods: the "bath," that is, immersion in the electrical atmosphere; the drawing of sparks from the electrified inert limbs; and the discharge of an electrical shock from the jar (fig. 1.7a, b; see Mauduyt 1784). The results obtained did not, however, live up to expectations:

> Although the electrification generated none of the effects we were chiefly expecting, those it produced at the outset and the genuine cures obtained elsewhere by the same means are such that no reasonable person can see any point in defending another opinion than this: electricity perseveringly employed and skillfully administered can be a useful remedy for the paralysis located in nerves or muscles. (Nollet 1749a, 407)

This was the talk of a cautious scientist sticking to the facts, albeit somewhat optimistically. But even the wariest could not ignore the reports of miracles, and Nollet was no exception: "How could I not be extremely surprised to see all these wonders confined, as it were, to a single country? . . . Such a legitimate astonishment aroused in me the inexpressible desire to see the facts and examine them from every angle—to understand, if possible, why they occurred only in Italy" (1749b, 446). On 27 April 1749 he was already on his way. When he reached Italy—from which he returned on 18 November—he proved to be the "scourge of all the charlatan physicists" (Mangin 1752, 22).

In all of the many cities he visited, he found evidence to form a very negative opinion of the renowned local electricians. In Turin, he called on Giovanbattista Bianchi, who administered the electrical purge in his presence even to Father Beccaria. But the intestines of this "thirty-five-year-old man of dry and bilious temperament" (Nollet 1749b, 447) proved impervious and the experiment failed. Doctor Bianchi muttered an excuse and the Abbé could not refrain from commenting that "most of the electrical cures in Turin have been mere fleeting shadows mistaken for reality with a bit too much haste or complacency" (ibid.). As for the purges, "the only thing to purge here is the imagination" (Nollet 1984, 176). More or less the same occurred in Venice. Here Pivati backtracked, denying consistent success with *intonacature* and finding excuses not to repeat the experiment. Then he really played the quack by promising the Abbé to turn up again and disappearing altogether. "He's a novice in physics," Nollet (1984, 177) wrote to Jallabert, "with little practice in experiments, something of a wonder-fancier and simpleton." In Bologna, things went even worse, if that were possible. Veratti proved to be anything but a "sober experimenter" (Beccaria 1758, 33). He confirmed everything—the cures and the electrical purges.

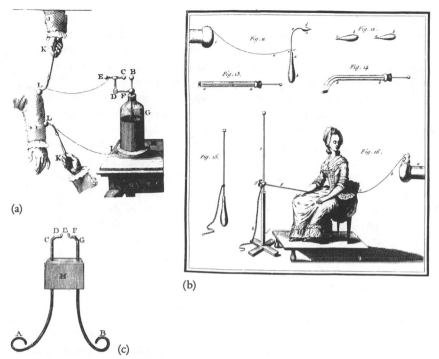

(a)

(b)

(c)

1.7 (a) An application of Leyden-jar electricity (from Cavallo 1781). (b) Electro-medical instruments and treatments. The figure shows the method of drawing sparks from the patient's electrified body by means of a grounded conductor (from Mauduyt 1784). (c) Electrical cures for toothache. "When the instrument must be used, it needs to be applied in such a manner that the afflicted tooth can be firmly secured in E by two threads, which, being flexible, can be arranged so as to hold teeth of different sizes; next, end A or B of one of the two irons must be connected by means of a metal chain or wire to the outer surface of a charged bottle, and the end of the other wire to the ball of the bottle, in such a manner that the shock travels through the instrument's metal wires and consequently across the tooth. A single shock thus discharged in a sick tooth will often cure it instantly; moreover, it is very advisable always to send two or three shocks through it" (Cavallo 1779, 368–69).

Nollet, who on this occasion was accompanied by a cardinal, asked Veratti to demonstrate his method on a young woman. The experiment failed but Veratti promptly produced an explanation worthy of a character in a Boccaccio tale: "Monsignore," he said to the cardinal, "it is because we could not electrify her in your presence as she should be electrified."[8]

The demonstrations fared no better elsewhere, and when Nollet returned to Paris, he was so disgusted that he seriously thought of abandoning the study of electricity. Probably to comfort him, Giovanni Fortunato Bianchini dedicated his *Saggio d'esperienze intorno la medicina elettrica* to Nollet. In the preface, we finally find the correct explanation of so many failures of electrical therapy—and, prophetically, of so many present-day psychoanalytic treatments as well: "Perhaps because in the act of electrifying ourselves, none of us had that firm conviction of having to feel some change from it—the conviction presumably held by those whose many wonderful achievements we hear about" (1749, xi).

In the meantime, however, work was going on outside Italy too. A favorite field was paralysis, regarded as the disorder most responsive to electrical treatment (Mangin 1752, 31).[9] Results in this area were obtained—or believed to be obtained—by Jallabert in Geneva (1748, 127–36; see Nollet 1984) and François Boissier Sauvages de la Croix and Jean-Etienne Deshais in Montpellier (see Mangin 1752, pt. iii). In Montpellier, the influx of patients was particularly strong. The populace, attracted by the whiff of magic, packed the hospital at an average rate of at least twenty people a day for two to three months. Then electrical medicine proliferated. The "divine and omnipotent virtue of electricity," as the Abbé Nicolas Bertholon called it, exerted itself everywhere. Aside from paralysis, which succumbed to "the triumph of electricity" (Bertholon 1780, 263), a few random examples culled from the plethoric literature of the period reveal that no ailment was spared: gout, irregular menstruation and amenorrhea, tertian fever, asphyxia, rheumatism, hysteria,[10] toothache (fig. 1.7c), headache, chilblains, mental disorders,[11] hemorrhages, diarrhea, deafness, blindness, and venereal disease. By the 1780s electricity had truly become a panacea and even a miracle, as in the case of the couple who, after ten years of infertility, "regained hope" through electricity—in Bertholon's words—thanks to a few turns of the crank and some shocks in the appropriate parts (the Abbé, demurely, did not specify which).

The more attentive and serious observers naturally counseled caution. A prescient Sguario had warned that the most one could concede to electrical medicine was an efficiency comparable "to an exercise, a stroll, or one of the better games in Gymnastics" (1746, 387). He was not alone in his opinion. Ten years later, Franklin wrote: "I never knew of any advantage from electricity in palsies that was permanent," adding wryly that "how far the apparent temporary advantage might arise from the exercise in the patients [sic] journey, and coming daily to my house, or from the spirits given by the hope of success, enabling them to exert more strength in moving their limbs, I will not pretend to say" (1941, 347).[12] Likewise, Musschenbroek (1768, 319) reported his failure to witness any positive results from electricity. And, when the fashion was raging, Tiberio Cavallo, author of a measured treatise on the subject (1781), warned against the "innumerable cures that purport to represent [electricity] as the panacea for every ill" (1779, 109; see also 1781, 3).

The point, as Sguario realized with his usual lucidity, was that electrical medicine depended on "experimental physics," precisely the area where it was quite weak. True, certain experimental facts were verified, such as the increase in fluid flow and perspiration (Nollet 1748a), the quickening of the pulse (Sguario 1746, 384), convulsions, glandular secretions, muscular contractions, and the evaporation of spirits. From these and other data one could—generally by analogy— draw conclusions about the human body. But apart from the fact that some of these findings were legitimately challenged from time to time, there was no pathological theory capable of acting as a framework for them and therefore of explaining the success or failure of a therapy. Worse still, no one felt this absence as a serious lacuna. The decay of empiricist philosophy, the preference for the rhetoric rather than the method of experiment, and the cult of confirmation rather than testing offered researchers an epistemological cover. For in practice they proceeded empirically, gathering every fact, classifying, measuring, and correlating every phenomenon. What little theory there was rested entirely on clinical cases, and, obviously, those that proved or were presumed successful confirmed the theory. In other words, electrical medicine underwent the same process as psychoanalysis in our time: in the long run, the prevailing concern with therapeutic aspects and the nearly exclusive use of the clinical method sterilized the theory. We need only glance at the innumerable texts and treatises of the period to find a plethora of confirmations of the "wonderful" virtues of electrical medicine; but hardly ever did a true critical spirit emerge to temper the enthusiasm.

Eventually, a few theories were introduced, but these were improvements, rarely advances. By the late 1770s, two schools of electropathology seemed to be taking shape. The first was what we might call the school of "electrical imbalance," chiefly represented by Pierre-Jean-Etienne Mauduyt de La Varenne and Bertholon. Mauduyt saw a correlation between health and atmospheric electricity: "Vigor and health consist in the equilibrium of vital forces. The alteration they experience in the second part (the negative one) of the storm seems a sufficient reason for the weakness and faintness felt by men and also by animals" (1776, 511). Bertholon elaborated the concept more systematically. For him, every animal body had its natural dose of electrical fluid. Illness was due to the disruption of this balance through an excess or deficiency induced by "imparted" electricity.

We can readily grasp the degree of empirical support for such a pathological theory. Health was defined in terms of electrical balance, and, if someone wanted to know when that balance occurred, the answer was quite simply: "If all actions are properly performed and all functions take place perfectly, we can be sure that the quantity of electricity in the human body at present exhibits the right balance required" (Bertholon 1780, 97–98). Bertholon did not seem greatly worried by this patent violation of the principle of noncircularity of explanations. Indeed, his entire work is so informed by this violation that it deserves to be read and meditated upon as an example of the ultimate aberrations induced by the pseudoscientific attitude of someone seeking to prove an *idée fixe* by force, even if in good faith. In this, Bertholon was truly behaving like a modern psychoanalyst: he was ever ready to suspect, and naturally to discover, an electrical impulse behind every phenomenon. It is impossible to find the shadow of a doubt in his manner of thinking: "Since all—or, if we prefer, nearly all—illnesses depend more or less on the electricity of the human body, it is obvious that electricity in general, used positively or negatively as circumstances require, is the means to cure them" (pp. 365–66). Hence: "If electricity sometimes fails to cure the illnesses for which its use was believed to be indicated, this may be due to the shoddiness of the method employed or to the impatience of the individual and of the electricians who are too quickly disgruntled." Exactly as in psychoanalysis, "success requires much perseverance" (369).[13]

The other electropathological school was rather better—for its greater moderacy if for nothing else—even though it boasted among its supporters not only a serious Tiberio Cavallo (1781, 11–13; 1782, 84–85), but also Nicolas-Philippe Le Dru, whom Volta described as a "renowned physicist juggler" (VO, 1:19) and "charlatan" (VO, 4:470). The basic idea of what we could call the school of "interruption of electrical fluid" was that illness is due to (a) a blocking of the course of the required vital fluid, (b) a contortion of the nerves, or (c) an increase in the fluid's viscosity. It was therefore thought that electrical therapy would prove effective in the case of (b)—which included paralysis—because electrical fluid, being very subtle, could clean the nerve canals where an obstruction or retention occurred.

But the theory's most distinguished and scrupulous advocate was unquestionably Luigi Galvani. His contribution to medical electricity remains little known and, with a few exceptions (Benassi 1963), is usually overlooked. Yet, at least methodologically, it was remarkable (for, as regards therapy, he did not depart from the prevailing practices). At any rate, his theory of animal electricity—a discovery that, as we shall see, had a far more solid experimental basis than Bertholon's—was not introduced ad hoc to support a therapeutic practice. It is true that, like Bertholon and Cavallo, whom he explicitly cited, Galvani too believed that certain illnesses "fluidum electricum indicant et . . . aucupant" (are evidence of and . . . entrap the electrical fluid), as he put it in an anatomy lesson of 27 February 1786 (1966, 120). Yet Galvani, like Sguario before him, displayed a clear awareness that medicine was a practice dependent on theoretical knowledge, particularly of a physiological nature, and that its efficiency was contingent on the existence of an independent basis for that knowledge. Indeed, it was only "after the origin of paralysis, and of natural contractions, as well as those arising from a diseased condition, had been set forth, (this was based principally on the discovered nature of animal electricity)" that Galvani allowed himself to discuss "several things . . . regarding the curing of these ills" (GM, 183; GOS, 312; GF, 85). All his major works follow this pattern, the theoretical-experimental part supporting the practical applications. Also, Galvani did not arbitrarily generalize electrical medicine to all illnesses, but confined himself to those best explained by the theory—namely, paralyses. Even here, he showed great caution:

> As to the treatment of paralysis, I see a problem full of difficulty and danger; for it is not easy to diagnose whether the disease originates from damage to the structure of the nerves or cerebrum, or from the presence of a non-conducting substance in the inner parts of the nerve or in other bodily parts wherein we believe an electric circuit is completed. If the first, artificial electricity, in whatever way it is employed, can be of little use, and perhaps can even be injurious; if it is the other, it seems possible some benefits can be derived from dissipating the non-conducting substance or by increasing the strength of animal electricity.
>
> At some future time, perhaps, experiment and experience will clarify the whole problem. (GM, 189; GOS, 315–16; GF, 87)

As for the physiological theory, it was based on a mechanism for the circulation of electrical fluid that Galvani himself called "conjectural" and that we shall examine later. The pathological version of such a theory employed other ancillary hypotheses. In normal conditions, it was argued, the nervous fluid was restored to the muscles by the arc of nerves. In the case of illness, the fluid would instead be either irritated by the stagnation of more or less conducting "acid humors," as in tetanus, or else blocked, as in paralysis (GM, 178–82; GOS, 309–12; GF, 81–83). For this theory Galvani relied on the analogy with the action of the artificial electrical fluid on animal nerves. This was, as we shall see, a typical approach of his—and indeed perhaps the best possible one given the scarcity of data available. But however prudent, not even Galvani's analogies sufficed to provide a solid theory. In such a situation, electrical medicine was fated to remain an orphan. In the event, it survived but, not surprisingly, it failed to make what Volta called "the advances that it seemed to promise" (VO, 1:49).[14]

1.3 THE LAWS OF ELECTRICITY

It has been said of Benjamin Franklin—who, in Turgot's famous epigram, "snatched lightning from the sky and the scepter from the tyrant" (see Franklin 1945, 502)—that "he found electricity a curiosity and left it a science" (Van Doren 1938, 171). While this verdict pays deserved tribute to Franklin's genius, it does not do full justice to history. However, the time span within which electricity emerged as an autonomous tradition of scientific research is very brief: no more than three decades, and possibly not even that, if we take Gray's first experiments as the conventional but not inappropriate starting point. Indeed, by the late 1760s—when Volta had completed his scientific training and Galvani, eight years his elder, gave his first public lectures in anatomy at the Archiginnasio of Bologna—some of the major discoveries in electricity had already occurred. These are worth examining because they constitute Galvani and Volta's common frame of reference—one could say the paradigm, if the technical meaning of the term did not make it highly unsuitable here. In discussing these discoveries, we ought to follow eighteenth-century usage and distinguish between laws and theories, that is, separate the facts from opinions. As Jean-Louis Alibert stated much later, but voicing a feeling already common in the period we are studying, "in physical science facts are undoubtedly immutable; but the hypotheses employed to explain their origin vary to infinity and in proportion to the imagination of the men who conceive them" (1802a, 77n).

The contemporary manuals and histories—and even those that appeared some time afterward, allowing for the inevitable lag in such works—are quite adequate for our reconstruction purposes. They also document this widespread epistemological belief. We shall draw on Joseph Priestley's *History and Present State of Electricity*, which originally appeared in 1767 and deservedly became a bestseller, often reprinted and translated; and Tiberio Cavallo's *Complete Treatise on Electricity*, published in 1777, which was equally successful and went through several editions and translations.

Let us begin, then, with the facts. The main properties of electricity known in the 1760s can be condensed into a small group of laws. Arranged schematically, they also provide a brief history of the first phase of the discipline and, more important for our purposes, an inventory of the experimental and theoretical knowledge that formed the starting point for Volta and Galvani. First, we must posit the

Laws of Electrics and Conductors

1. "All known substances are distributed by electricians into two sorts. Those of one sort are termed *electrics*, or *non-conductors*; and those of the other *non-electrics*, or *conductors*" (Priestley 1775b, 2:3; Cavallo 1782, 5).

2. "It is the property of all kinds of electrics, that when they are rubbed by bodies differing from themselves (in roughness or smoothness chiefly) to attract light bodies of all kinds which are presented to them; to exhibit an appearance of light (which is very visible in the dark) attended with a snapping noise, upon the approach of any conductor; and, if the nostrils be presented, they are affected with a smell like that of phosphorus" (Priestley 1775b, 2:4; Cavallo 1782, 1–2).

3. "When insulated bodies have been attracted by, and brought into contact with any excited electric, they begin to be repelled by it, and also to repel one another" (Priestley 1775b, 2:4–5).

4. "If conductors be insulated, electric powers may be *communicated* to them by the approach of excited electrics" (Priestley 1775b, 2:5).

Some of these properties are lost in the night of time. In particular, those mentioned in (2) were identified by William Gilbert and Otto von Guericke; those in (3) by Niccolò Cabeo and Guericke; those in (1) and (4) by Gray, who noticed that certain substances imparted "Electrick Vertue" and others not, and that this depended on their nature. One day, Gray was carrying out experiments on electrical transmission over long distances using a rope suspended from silk threads. An incident occurred that led him to his discovery. When one of the threads holding the rope broke, he replaced it with another, which, although equally thin, was made of brass. In this manner, however, the electricity traveled no farther. From this Gray concluded that "the Success we had before, depended upon the Lines that supported the Lines of Communication, being Silk, and not upon their being small" (Gray 1731, 29). The discovery of this property was followed by that of others—in particular, shortly afterward, the properties constituting the

Laws of the Two Electricities

5. "When two different bodies, except they are both Conductors, are rubbed together, they will both (provided that which is a Conductor be insulated) appear electrified, and possessed of different Electricities" (Cavallo 1782, 99; Priestley 1775b, 2:6).

6. "Bodies, possessed of the same Electricity, repel each other; but bodies, possessed of different Electricities, attract each other" (Cavallo 1782, 100; Priestley 1775b, 2:6).

Both discoveries occurred in continental Europe. They were made by Du Fay in 1733. Like Gray, the French scientist stumbled on them by chance. Du Fay was repeating Guericke's experiment on the repulsion of a thin gold leaf suspended in the air from an electrified glass tube. If another electrified body were brought up to the apparatus in this situation, one would reasonably expect a new repulsion. Such was indeed the case with rubbed rock crystal. But when rubbed amber and sealing wax were presented, they attracted the gold leaf instead of repelling it. The phenomenon, which repeated itself several times, induced Gray to formulate the "bold hypothesis" (*HAS*, 1734, 3) of two kinds of electricity. "This Principle," Du Fay wrote, "is that there are two distinct electricities, very different from one another; one of which I call *vitreous Electricity*, and the other *resinous Electricity*. . . . The Characteristick of these two Electricities is, that a Body of the *vitreous Electricity*, for Example, repels all such as are of the same Electricity; and on the contrary, attracts those of the *resinous Electricity*" (1734, 263–64).

Just over a decade later, the two electricities became, more precisely, two modes of being—positive and negative—of a single electricity. In 1747, a few months after receiving a gift of a glass tube from Peter Collinson, Franklin performed four famous experiments (1941, 174–75):

a. A person standing on wax, and rubbing the tube, and another person on wax drawing the fire, they will both of them, (provided they do not stand so as to touch one another) appear to be electrised, to a person standing on the floor; that is, he will perceive a spark on approaching each of them with his knuckle.

b. But if the persons on wax touch each other during the exciting of the tube, neither of them will appear to be electrised.

c. If they touch one another after exciting the tube, and drawing the fire as aforesaid, there will be a stronger spark between them than was between either of them and the person on the floor.

d. After such strong spark, neither of them discover any electricity.

From this Franklin concluded that "the Electric Fire is . . . not *created* by the Friction, but *collected* only" (quoted in Cohen 1941a, 66). "*A*, who stands on wax, and rubs the tube, collects the electrical fire from himself into the glass; and his communication with the common stock being cut off by the wax, his body is not again immediately supplied. *B*, (who stands on wax likewise) passing his knuckle along near the tube, receives the fire which was collected by the glass from *A*; and his communication with the common stock being likewise cut off, he retains the additional quantity received." Hence "we say B . . . is electrised *positively*; A, *negatively*" (Franklin 1941, 175).

During the same period, Franklin and his associates made another discovery, the

Law of Points

7. "If an insulated conductor be pointed, or if a pointed conductor communicating with the earth be held pretty near it, little or no electric appearance will be exhibited; only a light will appear at each of the points, during the act of excitation, and a current of air will be sensible from off them both" (Priestley 1775b, 2:6).

This "wonderful effect" was confirmed by two experiments. In the first, a sharp, thin stiletto tip removed the repulsion imparted to a cork ball by an iron shot placed on the neck of a charged Leyden jar—a sign that the point *attracted* the electricity of the iron shot. The second experiment consisted of applying "the point of a long, slender, sharp bodkin" to the iron shot. In this arrangement, the shot could not be sufficiently electrified to repel the cork—a sign that the point *repelled* the electricity of the ball (Franklin 1941, 171–73).

But Franklin's name is also linked to further, more momentous discoveries. Proceeding in chronological order, we find the

Law of Atmospheric Electricity

8. "A considerable quantity of Electricity exists in the atmosphere, and is certainly employed for some great purposes of nature" (Cavallo 1782, 101; Priestley 1775b, 2:10).

Lightning was unquestionably the chief natural phenomenon alluded to. The notion of its electrical nature must not have been hard to conceive after Guericke observed the sparks from the electrified globe and the typical small crackles

1.8 The Marly experiment. From Figuier 1868.

that accompanied them. Indeed, the idea was advanced by a number of scientists, including Hauksbee and Newton. In 1746 it was taken up by at least three authors: John Freke, Winkler, and Sguario (1746, 379). Two years later, Nollet noted that lightning and electricity exhibit analogies that made it possible to formulate "ideas sounder and more plausible than those hitherto imagined" (1748b, 314). But despite the controversy that ensued around Nollet (1774, letter vii),[15] it was Franklin who was responsible for the discovery, because he was the first to show how to verify it. In 1749 Franklin noted twelve properties common to lightning and electricity and, by analogy, inferred a thirteenth, namely, that lightning too was attracted by points. He therefore concluded: "Let the experiment be made" (1941, 334). His wish was fulfilled exactly in the manner he described, and the results matched his prediction (fig. 1.8). The experiment took place in France, at Marly-la-Ville, near Paris, on 10 May 1752, a Wednesday, at exactly 2:20 in the afternoon. Thomas-François Dalibard, in the company of the Comte de Buffon and Delor, had decided to test Franklin's hypothesis. In a garden in Marly, he planted a sharpened iron pole about one inch in diameter and forty feet high, resting on a three-legged stool supported by three wine bottles. Suddenly, while Dalibard was away, a thunderclap was heard. Coiffier, an old dragoon left on watch duty, immediately walked up to the pole with a Leyden jar and drew off two sparks, the second stronger and louder. The dragoon sent for the local priest. In the meantime, despite the hail, the experiment had attracted a large crowd. Seeing the lightning and the priest's hasty arrival, they concluded poor Coiffier had died. But, miraculously (not an inappropriate word, given the risk involved), that was not the case. The priest himself succeeded in drawing sparks from the pole until the storm ceased (see Dalibard in Franklin 1941, 257–62).

The Marly experiment was often repeated (in Italy, by Beccaria in Turin and by Veratti in Bologna; see Tabarroni 1966). Its impact on the popular imagination was enormous. But, from the standpoint of scientific knowledge, its importance certainly did not match the discovery of two other laws of artificial electricity. These played a prominent part in Galvani's first experiments and Volta's explanation of the latter. They are the

Laws of Electrical Atmospheres

9. "If a conductor, not insulated, be brought within the *atmosphere*, that is, the sphere of action, of any electrified body, it acquires the electricity opposite to that of the electrified body; and, the nearer it is brought, the stronger opposite electricity doth it acquire, till the one receive a spark from the other, and then the electricity of both will be discharged" (Priestley 1775b, 2:7; Cavallo 1782, 100–1).

10. "If this Conductor does not communicate with the earth, but is insulated, and approached to the excited Electric as before, then not only the side of it which is towards the Electric, but the opposite one also, appear electrified; with this difference, however, that the side, which is exposed to the influence of the Electric, has acquired an Electricity contrary to that of the excited Electric, and the opposite side an Electricity of the same kind with that of the Electric" (Cavallo 1782, 36–37).

The discovery of the laws of electrostatic induction, expressed in these laws of "atmospheres," began with Franklin's analysis of the Leyden jar. His contribution is very important for our purposes because the principle of the Leyden jar was crucial to Galvani's hypothesis of animal electricity. Franklin gave his interpretation in 1747–48 in two letters to Peter Collinson, where he laid down three principles:

1. "At the same time that the wire and top of the bottle, &c. is electrised *positively* or *plus*, the bottom of the bottle is electrised *negatively* or *minus*" (Franklin 1941, 180).

2. "The two states of Electricity, the *plus* and the *minus*," are "wonderfully . . . combined and balanced in this miraculous bottle" (p. 181).

3. "The whole force of the bottle, and the power of giving a shock, is in the GLASS ITSELF; the non-electrics in contact with the two surfaces, serving only to *give* and *receive* to and from the several parts of the glass" (p. 191).

Franklin obtained confirmation of the first two principles with four famous experiments that showed the usefulness and explanatory capacity of his concepts

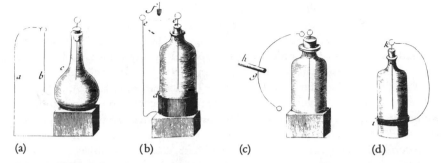

(a)   (b)   (c)   (d)

1.9 Franklin's analysis of the Leyden jar. (a) A linen thread hanging from a wire is at-tracted by the charged bottle every time one touches the top of the bottle: "As soon as you draw any fire out from the upper part . . . the lower part of the bottle draws an equal quantity in by the thread." (b) A small cork ball hanging from a silk thread "plays inces-santly" from the top of the bottle to the ringed end of a wire joined to the bottle's outside coating; the ball therefore "fetches and carries fire from the top to the bottom of the bottle, 'till the equilibrium is restored." (c) If, using a metal wire held by a sealing-wax handle, one touches the bottom of the electrified and insulated bottle first, and then gradually brings the device close to the top, one will draw sparks. The same happens if one touches the top first, then the bottom. If one touches the top and bottom simultane-ously, the jar will discharge instantly: this demonstrates the gradual or instant flow of electrical fluid "till the equilibrium is restored." (d) A metal wire connecting a lead or paper ring to the top of the bottle prevents it from charging. In this case "the equilibrium is never destroyed: for while the communication between the upper and lower parts of the bottle is continued by the outside wire, the fire only circulates: what is driven out at bottom, is constantly supplied from the top." From Franklin 1941, 162 (ill.), 182–83 (text).

of positive and negative electricity (fig. 1.9). In a series of equally celebrated experiments, he proved the third principle, namely, that the charge resided in the glass. This was shown by the following facts: (1) The shock was also felt when the cork and wire were removed from the electrified bottle; (2) no shock was felt when the water was decanted into an empty bottle and touched; (3) the shock was felt once again when the first bottle was filled with fresh water (pp. 191–92).

From these experiments, Franklin concluded that "the equilibrium cannot be restored in the bottle by *inward* communication or contact of the parts; but it must be done by a communication formed *without* the bottle between the top and bottom, by some non-electric, touching or approaching both at the same time," that is, the two coatings separately (p. 180). He also concluded that "in a bottle not yet electrised," no electrical fire "can be thrown into the top, when none *can* get out at the bottom" (p. 181). As for the glass, it carried a maximum of electricity, Franklin argued, and it would not support any increase; but in the charged bottle, its electricity was unbalanced, as it carried more on one side than on the other. The equilibrium, however, could not be restored through the glass itself, which was impermeable, but through the outside, by putting the two sides in contact via a conductor. Lastly, the shape of the glass was irrelevant: the cylindrical shape of the bottle could be replaced by any plate armed on both sides with metal foil. This was the "Franklin square" or plate condenser.

Franklin's analysis of the workings of the Leyden jar led to law (9), because it showed that a conductor electrified through contact induces on another, grounded conductor an electricity of opposite sign when the two are separated by an electric (the glass). Law (10) stems from the analysis by Aepinus of John Canton's celebrated experiments. In one of these, Canton had observed that two small cork balls hanging from an electrified tin tube drew apart; that, "at the approach of the excited [glass] tube, they will, by degrees, lose their repelling power, and come into contact; and as the tube is brought still nearer, they will separate again to as great a distance as before" (1753, 294). Conversely, when the glass tube was slowly removed, the cork balls "will approach each other till they touch, and then repel as at first." Aepinus saw in these experiments the same mechanism as the Leyden jar: two surfaces—the excited glass tube and the tin tube—inducing opposite charges in each other. There was a difference in that one of the surfaces was not grounded. But another, more important difference was the type of electric interposed. In Canton's experiment, it was air. Impermeability, therefore, was not—as Franklin claimed—a property of glass, but of all electrics (Aepinus 1759, 10). Aepinus fully confirmed this assertion by constructing the parallel air condenser (1756, 119–21).

1.4 THE "SUBLIME THEORY" AND ITS PROBLEMS

In his *Treatise*, Cavallo wrote:

> It is the business of Philosophy to collect the history of appearances, and from these to deduce such mechanical laws, as may either be themselves of immediate use, or lead to the discovery of other facts more interesting and necessary for the happiness of human kind. After a number of such constant appearances, which are called natural laws, have been established and confirmed by a sufficient quantity of experiments, it is then proper to investigate the cause of these effects; which, if it is once discovered, and its mode of activity is ascertained, puts an end to the trouble of experimental investigation, and renders the application of its effects certain, and determinate. (1782, 103)

In this case, laws were indeed followed by theories, two of which are particularly notable. The first was advanced by the Abbé Nollet. Electricity, he argued in substance, is a highly subtle fluid that penetrates all bodies with two opposite motions: *affluent* and *effluent*. The penetration occurs with varying degrees of

1.10 Nollet's theory. The effluent matter that emanates in divergent rays from the electrified body is gradually and simultaneously replaced by the affluent matter that arrives in convergent rays. Corpuscle G, subjected to both currents of fluid, is driven toward the electrified body by the affluent matter, which is the stronger one here. From Nollet 1747.

facility. The two currents of fluid travel in straight jets—divergent for the effluent and convergent for the affluent. Moreover, the pores of the effluent rays differ from those of the affluent rays, and in some bodies the effluent current is stronger than the affluent one (fig. 1.10). The other theory belongs to Franklin. Initially, it rested on the following principles (1941, 213–14, 216):

1. "The electrical matter consists of particles extremely subtile."

2. "Electrical matter differs from common matter in this, that the parts of the latter mutually attract, those of the former mutually repel each other."

3. The particles of electrical matter exhibit a "strong attraction between them and other matter"; thus "common matter is a kind of sponge to the electrical fluid."

4. In common matter, electrical matter is in equilibrium. If more electrical matter is added, "it lies upon the surface and forms what we call an electrical atmosphere."

5. "The form of the electrical atmosphere is that of the body it surrounds."

6. "All kinds of common matter do not attract and retain the electrical, with the same strength and force . . . those called electrics *per se*, as glass, &c. attract and retain it strongest, and contain the greatest quantity."

After Canton's experiments, Franklin successively added two other principles (p. 302):

7. "Electric atmospheres, that flow round non-electric bodies, being brought near each other, do not readily mix and unite into one atmosphere, but remain separate, and repel each other."

8. "An electric atmosphere not only repels another electric atmosphere, but will also repel the electric matter contained in the substance of a body approaching it; and, without joining or mixing with it, force it to other parts of the body that contained it."[16]

Franklin's theory was accepted, at least in substance, by Jean-Baptiste Le Roy and later Sigaud in France, Watson and Priestley in England, Johan Carl Wilcke and Aepinus in Germany and Russia, and Beccaria and Volta in Italy—among others. The reasons for its rapid success pose a singular challenge to those who believe that the history of science exhibits a single criterion for theory choice and for progress. Comparing Franklin's theory with Nollet's, Franklin's French translator Jacques Barbeu Dubourg attributed the superiority of the former to two reasons that read as if they had been lifted, two centuries ahead of time, from Imre Lakatos's methodology of scientific research programs. Barbeu wrote:

Here is the principle that I regard as the required starting point. Of two hypotheses, the better one, in my view, is the one that, encompassing all the known facts and highlighting the advantages and differences, links them together in an order so beautiful that we can not only grasp all their relationships at once, but also, almost at a glance, perceive what they lack in completeness and what remains to be done to fill the gaps or add scattered links in the great chain of physical truths.

If, on this basis, we wish to assess Nollet's hypothesis of affluences and effluences and Franklin's hypothesis of positive and negative electricity, I don't believe we can remain in doubt for long.

Nollet interprets everything in a vague and indistinct manner in terms of affluences and effluences. He does not teach us to understand, even less to predict anything. He offers a stitch for tying together facts that are known or still to be discovered, but he provides no thread for getting out of the labyrinth in which we must seek them. It is as if a botanist simply showed us that trees have a trunk, roots, branches, leaves, flowers and fruit, but without telling us either which features we can use to distinguish a given tree or what constitutes their distinctive character. Yet such information is much more valuable than the laborious repetition of the same general statements for each objet.

Franklin, instead, distinguishes an electricity that is sometimes positive, sometimes negative. He assigns to each its proper place and character—as far as the present state of knowledge in physics allows. In so doing, he spreads the light up close and afar, showing the path we must follow to make new discoveries, to relate them to earlier discoveries, to drive back the limits of science and enable us find not only pleasure but also a manifest utility in it. Franklin says: *do this, and here is what will happen; change this circumstance, and here is what will result; arrange things thus and you will avoid such and such inconvenience.* We follow his texts and everything occurs in the manner and order he announced, everything squares with his views, in Europe as in America; everything, even celestial phenomena, shows the solidity of the principles that modesty has not allowed him to put forward otherwise than as mere conjectures. In short, I believe that you, like me, will find that between the theories of these two famous electricians there is just about the same difference as between a barren fig-tree and one that bears fruit. (1773, 336–37)

Perhaps it is for these reasons that Sigaud (1785, 146, 246, 571) repeatedly referred to Franklin's theory as "sublime." But in fact the situation was not as clear as the enthusiastic Barbeu made it out to be. Nollet's theory admittedly suffered from major gaps, in particular as regards the explanation of the Leyden jar, where Franklin's theory exhibited its "chief excellence" and unquestionably "that which gave it the greatest reputation" (Priestley 1775b, 2:29). For the rest, however, Franklin's theory too raised many unsolved problems. Even its predictive power, which Barbeu regarded as its chief virtue, was in reality an ex post facto arrangement using ad hoc hypotheses totally inadequate to explain new phenomena. Examples of this include the two principles of atmospheres introduced in order to explain Canton's experiments. But, above all, the theory ran into trouble with the most basic electrical phenomena: it explained the dispersive power of points (fig. 1.11) but not their attractive power; it explained the attraction between positive and neutral bodies, the attraction and spark between positive and negative bodies, the repulsion between positive bodies or between a positive and a negative body, but it failed to explain the repulsion between negative bodies or between a negative and a neutral body.[17] By contrast, some of these phenomena were fairly well accounted for by Nollet's theory.

1.11 The dispersive power of points according to Franklin. The electrical atmospheres *HABI* and *KBCL* are held by surfaces of matter (respectively lines *AB* and *BC*) far larger than those holding atmospheres *IBK* and *LCM* (respectively points *B* and *C*). From Franklin 1941.

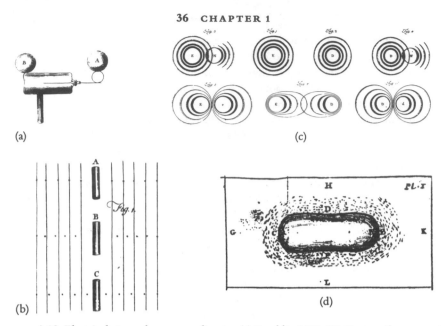

1.12 Electrical atmospheres according to: (a) Franklin 1941; (b) Canton (from Priestley 1775b); (c) Beccaria 1772; (d) Stanhope 1779.

The most awkward problems stemmed from the concept of "electrical atmosphere," introduced by Franklin to explain attractions and repulsions (fig. 1.12a). Such atmospheres, portrayed as actual clouds of material surrounding the electrified body, were supposed to act through contact. But as Roderick Home has impeccably shown (1972; 1981, chap. 7), this put Franklin's theory in great jeopardy. For example, why didn't the electrical atmospheres, which were fluids of a certain kind, mingle with each other and with a body's electrical fluid? Franklin simply turned these difficulties into axioms of the theory, as he needed them to explain the laws of induction. But with this move, his already contrived theory became altogether indefensible in the face of Aepinus's air condenser. If Franklin's theory applied, the device could never have become charged, because the atmosphere of one plate would have invested the other and thus discharged the electrical fluid into the ground.

This stumbling block marked the start of the agonizing attempts to revise Franklin's theory (see Heilbron 1971), which was shortly transformed into at least two versions (see Home 1972, 1979). The first was developed by Canton and Beccaria, who tried to explain electrical signs by means of fluid flux. The second was also due to Canton and Beccaria as well as to Lord Mahon (Charles Stanhope), all of whom introduced roughly similar versions of a theory on the excitation of the air's electrical state (fig. 1.12b, c, d). Finally, there was a total change of course. First Aepinus, then, independently, Volta and Cavendish completely rejected the notion of material electrical atmospheres and espoused that of action at a distance. Beccaria's intellectual itinerary is especially significant because his views and celebrated tests influenced even the physiologists who, like Galvani, had resorted to electricity to explain muscular movement.

Beccaria began by arguing that "every electrical sign is due to the vapor that expands from the body where it exists in a major quantity into the body where it exists in a minor quantity" (1753, 17). He was, however, incapable of explaining the "individual mechanical manner" in which this occurred.[18] Subsequently, he decided to introduce what he called a "hypothesis or, we should say, fable" but also a "plausible conclusion": when a more electrified body A enters the atmosphere of a less electrified body B, the vapor that forms the atmosphere of A flows into that of B. The flow shifts the air and creates a vacuum, "whence the two bodies A and B, pressed by the air on their backs and less pressed (because of the ejected air) on their fronts will eventually join together" (1758, 43). Finally, he abandoned even this notion and claimed that "in essence, the electricity of a body A does not diffuse into the surrounding air, in other words, if a body A is electrified in excess, the added fire does not penetrate, at least to a perceptible degree, into the substance of the surrounding air" (1772, 173; 1769b).

Although this new model described the electrical atmosphere no longer as a cloud of vapor external to the body, but as an excited state of the air's electrical fluid, the motions were still explained in terms of contact action. This did not solve all the difficulties, and we can understand why Aepinus—followed independently by Volta and Henry Cavendish—switched to a theory of forces acting at a distance. The electrical atmospheres did remain as "spheres of action," but the physical referent disappeared. This marked a major theoretical innovation. Volta started there and continued along his own path; others, either because they were more conservative or, like Galvani, not directly engaged in leading-edge physical research, preferred to abide by the traditional concept. We shall see how this circumstance too, at least initially, influenced the controversy regarding animal electricity.

CHAPTER 2.

Volta's and Galvani's Scientific Training

2.1 SYMMER'S EXPERIMENTS AND VOLTA'S ELECTRICAL THEORY

On 1 February 1759, Robert Symmer, paymaster to the treasurer of the king of England's Chamber and a quite talented scientist, delivered his first paper to the Royal Society, of which he had been a fellow since 1752. Symmer reported a new experiment that, although inelegant, seemed in his prophetic words "to open a new path for proceeding in electrical researches" (1759, 341). The scientific novelty of the experiment resided in the fact that two insulating bodies, when electrified in opposite ways, lost all sign of electricity when brought joined together, but regained their electrical charge when drawn apart. The inelegant feature was that the bodies exhibiting such odd behavior were Symmer's stockings. In winter, Symmer usually wore two pairs, one of black wool, the other of white silk. In the morning, he would carry the silk pair over the woolen pair; in the afternoon, vice versa. When he took off his stockings, he noticed the typical signs of electricity—crackles, tiny sparks, and a slight breeze on his hand. By repeating the experiment with greater care and attention, and in different circumstances, Symmer was able to observe the phenomena in detail. His most relevant findings were the following:

> Both the stockings, when held at a distance from one another, appear inflated to such a degree that, when highly electrified, they give the intire [sic] shape of the leg; and when brought near the face, or any naked part of the body, there is a sensation felt, as if a cool wind was blowing upon that part. When the two white, or the two black, are held together by the extremities, they repel one another, and form an angle, seemingly of 30 or 35 degrees.
>
> When a white and a black stocking are presented to each other, they mutually attract, with a force answerable to the degree of electricity they have acquired: when brought within the distance of three feet, they usually incline towards one another: within two and a half, or two feet, they catch hold of each other; and when brought nearer, they rush together with surprising violence. As they approach, their inflation gradually subsides; and their attraction of foreign objects diminishes: when they meet, they flatten, and join as close together, as if they were so many folds of silk; and then the balls of the electrometer are not affected at the distance of a foot,

nor even of a few inches at certain times. But what appears most extraordinary, is, that when they are separated, and removed at a sufficient distance from each other, their electricity does not appear to have been in the least impaired by the shock they had in meeting. They are again inflated, again attract, and repel, and are as ready to rush together as before. When this experiment is performed with two black stockings in one hand, and two white in the other, it exhibits a very curious spectacle: The repulsion of those of the same colour, and the attraction of those of different colours, throws them into an agitation that is not unentertaining, and makes them catch each at that of its opposite colour, at a greater distance than one would expect. When allowed to come together, they all unite into one mass; when separated, they resume their former appearance, and admit of the repetition of the experiment as often as you please; till their electricity, gradually wasting, stands in need of being recruited. (Pp. 353–55; see fig. 2.1)

Symmer's experiment—which made him known as the *philosophe déchaussé* (barefoot philosopher)—was in itself a fairly common phenomenon, then as now. In Italy, it had already been performed by Alessandro Amedeo Vaudania and described by Beccaria (1753, 197). Subsequently repeated by Nollet, it was the focus of intense discussion because both Nollet and Symmer saw it as confirming the existence of two distinct electrical fluids, rather than the single fluid postulated by Franklin. In reality, the experiment was not so conclusive. In the

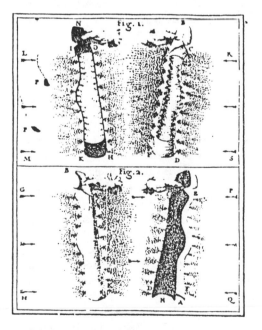

2.1 Symmer's stockings. From Nollet 1774.

ensuing debate, Priestley—whose radical empiricism did not prevent him from occasionally recognizing the "theory-ladenness" of observation—went so far as to qualify the experiment as "a remarkable instance of the power of an hypothesis in drawing facts to itself, in making proofs out of facts which are very ambiguous, and in making a person overlook those circumstances in an experiment which are unfavourable to his views" (Priestley 1775b, 1:322).

Priestley, however, was too harsh. While inconclusive for the two-fluid theory, Symmer's experiment clearly posed new problems for the Franklinists. The version performed in Turin by Giovan Francesco Cigna brought those difficulties into sharp relief. One of his experiments is especially noteworthy because—as Volta later remarked—it shows how Cigna "anticipated" the discovery of the electrophore (or electrophorus) (VO, 3:142). The experiment was the following:

> I suspended a light, flat plate of lead from silk threads to insulate it. I then presented to the flat surface of the lead a ribbon charged with vitreous electricity, holding it at one end with my hand away from the plate. If in the meantime I brought my finger up to the lead, a spark would spring between the lead and my finger; immediately the ribbon was drawn to the lead and, clinging fast to it, easily carried the full weight of the plate. The ribbon then continued to adhere to the lead and, although joined together, neither the lead nor the ribbon attached to it gave any further electrical sign. If the ribbon was detached from the lead, a new spark flashed between the lead and the finger presented to it, and the ribbon displayed electricity as before. (Cigna 1765, 43)

How could this phenomenon be reconciled with Franklin's theory, which held that the contrary electricities should have canceled each other out completely? Cigna believed he had found a way out of this dilemma by distinguishing between two kinds of electricity: the Symmerian, which dissipated slowly, and the Franklinian, which dissipated instantly.

Beccaria, for his part, thought he had found a more orthodox approach with respect to Franklin's theory by introducing the principle of "vindicating electricity" (1767, 1769a, b). Beccaria's thesis, as expressed in his most mature text on the subject, was as follows: "I. Electrified insulators lose their present electricity when they move into the adhesive state. II. When disjoined, the bodies recover it" (1772, 410). This "property" or "disposition" of insulators constituted the essence of "vindicating electricity."

But Beccaria's system was by no means the only alternative solution to the problem. In *Elettricismo artificiale*, he himself alluded to another theory—admittedly to refute it—according to which "the electrified insulators brought to the said adhesive state retain their existing electricities" (1772, 407). Beccaria did not even mention the name of the man who had introduced this theory back in 1769: Alessandro Volta. In his silence, Beccaria followed the practice that was—

and perhaps still is—customary in relations between illustrious masters and young unknowns. Volta was then just twenty-four. Born in Como on 18 February 1745, this self-taught young man, who suffered a difficult childhood (see Z. Volta 1879; A. Volta, Jr. 1900; Volpati 1927), possessed—as he himself put it—"a genius for electricity." This experimental and theoretical talent soon made him one of the leading scientists of the late eighteenth century. It was manifest from his very first printed work (see Massardi 1926), *De vi attractiva ignis electrici*, published as a letter to Beccaria dated 18 April 1769.

Even more important than its merit, the work's chief novelty lay in its method. Beccaria had formulated vindicating electricity as an *additional* principle to the Franklinian theory. On this point, he was explicit, and in a letter of 20 February 1767 indirectly defended himself against the obvious suspicion of having introduced an ad hoc hypothesis. Beccaria claimed his new principle could unify and explain otherwise unrelated facts: "Are they not to be regarded as having made a great advance in Physics, who devise experiments, the greatest in number and most diverse in appearance, and show from what one principle all of them depend, and that separate phenomena are bound with each other as if by a mutual relationship—[phenomena] which before appeared most independent and almost contrary?" (1767, 43 [Latin], 50 [English]).

Volta took the reverse approach by advocating the methodological principle of the parsimony of explanatory hypotheses. After his initial statement of it in 1769, the principle was to remain an abiding feature of his work, featuring prominently too in his later controversy with Galvani. Volta objected to Beccaria practically in the latter's own terms: "Will it not perhaps be a great advance for physics, when the principles of Franklinian theory—which are already few and simple in themselves, but require a further, simpler cause to which they can in turn relate—are reduced to the sole principle of attraction? When the Franklinian principles are reduced, I mean, together with those You thought proper to add?" (VO, 3:24–25; VOS, 51).

In essence, Volta was giving Franklin's beard a heavy shave with Occam's razor. The resulting theory was based on two fundamental principles that we can illustrate with quotations from Volta's *Saggio teorico e sperimentale di elettricità* (1778–80):

1. There is "an electrical fluid, copiously distributed in bodies to such a degree and in such a manner that all bodies possess it in a quantity commensurate with their capacity" (VO, 4:384).

2. There are "attractive forces of the smallest parts of every body" that balance the attractive force of the electrical fluid (VO, 4:381).

The first—the principle of "natural saturation"—was not new. But the second was Volta's own: at the time he wrote *De vi attractiva*, he was not yet acquainted with the work of Aepinus. In discussing this principle, Volta made it clear from the start that the force of attraction at a distance he had in mind was not the Newtonian gravitational force. In addition to the latter, which regulates the macrocosm, there were other, "nonmechanical" forces. This was proved by the refraction of light, the attraction between drops of water placed at a minimum distance from each other, and so on—the phenomena discussed by Newton himself in the last Query of the *Opticks*. Nor, Volta added, should we be frightened by this multiplication of forces. Indeed, "we can recognize that only two or three types of forces apply to the primordial particles." One could even admit, with Roger Joseph Boscovich, a single law of forces from which could be deduced "the other types of attraction in smaller bodies and over shorter distances" (VO, 3:26; VOS, 53). On the other hand, while we should not be surprised by the existence of an attractive force different from the gravitational one, we should not be disconcerted either by the fact that it operates at some distance. The electrical force is not the only force to behave thus: the magnet, for example, attracts in the same manner and even more so. And in this connection we should also bear in mind that

> it is not at all necessary that the attractive force of the electrical fire should exert itself over a distance as great as the interval between two mutually attracting bodies; for, since the surplus fire spills out to surround the electrified body, forming a sort of atmosphere, we can understand how a thread placed, say, two feet away from the chain is not equally distant from the outer limit of the chain's atmosphere; on the contrary, it already feels the attractive force when it comes close to that limit. (VO, 3:29; VOS, 58–59)

As we can see, Volta's theory combined new ideas with a residue of the old notion of material electrical atmospheres. But this vestige had an archeological flavor and seemed almost a deliberate inclusion to enable the recipient of the letter to digest a particularly distasteful concept for him. Indeed, the fact that Volta no longer attributed a major explanatory function to electrical atmospheres is also fairly clear from his objections to the contemporary versions of Franklin and Nollet's theories: "For the electrical motions are either caused by the pressure of some fluid, or have no other cause than the one mentioned, namely, the attractive force of electrical fire" (VO, 3:26; VOS, 54). The first alternative, Volta argued, was reduced to two hypotheses: (1) (first version of Franklin's theory, and Nollet's theory) the electrical fluid itself caused the motions: but this was impossible, because the fluid flowed in only one direction, as Beccaria himself had proved; (2) (second version of Franklin's theory, finally accepted even by Beccaria) the displaced air caused the motions: but even this was impossible because, Volta argued—capitalizing on an experiment by Cigna (1765)—"if the convergence of two bodies possessed of different electricities, when they are in the air, were due to the pressure of the air impelled into resuming its earlier position, the same phenomenon should occur when the bodies are immersed in oil, a similarly insulating medium" (VO, 3:27; VOS, 55). But that does not happen.

By contrast, attraction at a distance, together with the principle of the natural quantity of fluid, offered an easy explanation for the motions of bodies. The idea was the following. In normal conditions, there is an equilibrium or "saturation" between the fluid and the forces. As a result, there are no signs of electricity. But when, for any reason, an imbalance occurs—that is, an increase or decrease of forces—the unbalanced body, the sum of whose attractive forces is thus augmented or diminished, "craves" or "is craved" by new fire. Let us consider, for example, a body unbalanced through a deficiency of forces (that is, overelectrified). "The bodies around it, including the air, attract that body's excess fire by means of their own force, and are in turn attracted by that body." As the surplus, unsaturated fire is light and mobile, it would tend to flow out of the body into the others. However, "we must consider that the interposition of air, an insulating medium, prevents the fire from passing freely from the overcharged body to the other bodies some distance away; hence it is rather the latter, if they are light enough, that will move toward the electrified body." "This," Volta stated, "is the only law with which all electrical motions generally comply: indeed, even in bodies charged with identically named electricities, what do their mutual divergences show, if not the unequal attraction toward other bodies, including the air?" (VO, 3:28; VOS, 56–57).

With this theory, Volta also explained the Leyden jar, the Franklin square, electrostatic induction or "application," and the Symmerian phenomena. Here Volta's explanation and Beccaria's differed considerably. Let us consider a positively charged glass plate and a sheet of metal foil. Bring the foil into contact with the glass and touch it, drawing sparks. In these conditions, neither body displays electricity. According to Beccaria, this happens because the bodies "lose" their electricity when in contact and "recover" it when separated. On the contrary, according to Volta, both bodies keep their electricity. The glass keeps its positive charge, because "one must not think that all the electricity has left the glass—that is, that the surplus fire has been completely removed from it—for the glass could not have discharged its fire in so short a time" (VO, 3:46; VOS, 82). The foil keeps its negative charge, because part of the surplus fluid "applied" to it by the glass has been expelled in the form of sparks. Thus, when the bodies are newly drawn apart, they will each display their own electricity. At this point, one repeats the operation: the experimenter touches the foil again—restoring it to a state of natural saturation—and brings it up once more to the glass. If the foil is touched yet again, it will lose a further surplus of fluid "applied" by the glass, and the two bodies will cease to display electricity. Volta concluded that when the detached bodies exhibit electricity, it is because the glass "recovers the electricity it had before it was touched" and "not because the absolute electricity of the glass has been, so to speak, reconstituted as a result of the divestiture" (VO,

3:47; VOS, 83). In other words, Volta argued for the existence of a mere "vindicated electricity" (*electricitas vindicata*) rather than a "vindicating" one (*electricitas vindex*). Two principles are superfluous when one will do.

Volta still had to explain why and how the system of forces in an electrified body became unbalanced. The causes to which he ascribed the phenomenon were: rubbing, the commonest and most important; percussion and pressure, "which can in fact be equated with rubbing" (VO, 4:385); heat, as in the case of tourmaline; and, as we have seen, induction. As for the mechanism, the "specific mechanical manner," as Beccaria would have said, one had to suppose that "the mutual attractions depend on the mode and force of action of the arrangement, shape, etc., of the minutest surface parts of bodies." Rubbing and other actions alter the positions and therefore the sum of forces. The disturbed equilibrium will tend to be restored—"instantly" in conducting bodies and "slowly, gradually," in nonconducting ones. The change in balance must naturally occur even when the rubbed bodies are conductors, for example two metals—in which case, however, the balance is restored instantly. That is why, to observe electrical signs, at least one of the bodies must be a nonconductor, because on nonconductors the equilibrium is established slowly and gradually.

Are there other causes for the imbalance in the system of forces? It should be noted, in this connection, that Volta laid the premises for regarding contact as a borderline case of rubbing. *De vi attractiva* did not mention it, nor did the *Saggio* and *Lezioni compendiose sull'elettricità* (written around 1784), but the argument is explicitly stated in a manuscript note attached to a *Descrizione ed uso dell'elettroforo*, which, although undated, is certainly earlier than 1792. In the note, Volta discusses electrification by percussion, pressure, and heating. He describes some experiments conducted for the purpose of "determining the minimum touching and compression required to excite a sign of electricity even on glass," adding:

> I have performed the same experiment and with the same—indeed greater—success by placing the crystal pane on the mercury; every time I lifted it off it gave me very clear signs of electricity. Now from these experiments we can note that a small pressure will suffice to transmit the electrical fire from one body to the other; as the fire will consequently be unbalanced in both bodies, it will display its presence with the usual signs only when the bodies are drawn apart. I would not even wish to assert that pressure is required; perhaps *contact* alone is enough; and pressure intensifies the effect for no other reason than because it brings more points into contact: thus percussion intensifies it even more; as does *rubbing*, by multiplying the contacts between the surfaces almost infinitely. (VO, 3:175n)

This passage is an important statement of the notion that mere contact between two bodies generates an electrical imbalance—or "electromotive power," as Volta later called it. We shall see the decisive weight of this idea in his dispute with Galvani over animal electricity.

2.2 FROM THEORIES TO INSTRUMENTS:
VOLTA'S STRATEGY AND EPISTEMOLOGY

Volta emerged on the scientific scene through a controversy with Beccaria. That was a sign of fate. But another sign was the manner in which the debate was conducted and concluded. For Volta inaugurated a strategy and formed epistemological convictions to which he adhered for the rest of his career.

There are so many ways in which a confrontation between rival theories may be settled to the advantage of one of them. A crucial experiment may be found that will refute one of the theories; or one theory may find itself gradually overwhelmed by so many difficulties and anomalies as to disqualify itself, possibly after many attempted adjustments; or one theory may encounter so many successes—or a single success of such moment—as to make its failures pale in comparison with those of the other theory; or one theory may be judged to hold little promise. Volta blazed a new trail: the translation of a theory into an instrument. This was, and is, a practical means to resolve a theoretical conflict. If an instrument can be derived from a theory, and if that instrument works, then there is at least a specific reason to believe that the theory works.

In the study of electricity, Volta employed this strategy at least twice, on the occasion of two separate controversies. The first time was when he supported the theory of attractive forces and electrical atmospheres, which pitted him against not only Beccaria but also the orthodox and reformed Franklinists: this resulted in the invention of the electrophore and a family of other instruments. The second instance was when Volta advocated the theory of contact electricity between dissimilar conductors, against Galvani and the supporters of the animal-electricity theory: this produced the Voltaic pile.

From the theoretical standpoint, the strategy is far from reliable. That a theory gives rise to an instrument does not mean the instrument's *efficiency* proves the theory's *truth*. Viable instruments can be built on false theories, and there are a number of cases where instruments are used successfully despite the lack of a covering theory. In itself, an instrument is a fact that requires explanation. The theory underlying its construction does not always provide that explanation or at least a good explanation. An excellent example is the electrical instruments and cures that Volta himself treated with reservations and disparagement.

However, as regards practical results, the strategy of using an instrument to support a theory is certainly quite efficient. It forces the adversary onto the defensive by confronting him with an unexpected fact. It also generally wins support for the inventor of the instrument, not because the instrument actually validates his theory and falsifies his opponent's, but because the instrument causes attention to shift from the merits of the theory to technical performance. As a result, the success of such performance, or the enthusiasm for it, has a vicarious effect on the theory, enabling it to infiltrate almost surreptitiously, encircle the opposing theory, and drive the latter to spontaneous surrender. Volta himself described something similar to this drift effect when recounting the reception of his electrophore:

> Now as soon as my apparatus came on the scene, its effects—which were all the greater and more surprising as they were easily obtained—must have struck all observers and blinded them: the impressive name of *Perpetual electrophore* also helped increase that astonishment. Lastly, the passion for novelty and wonders led people to believe that everything was new and wonderful, so that by linking the invention of the name and apparatus to the invention of that type of electricity, everything was indiscriminately attributed to the same author. (VO, 3:137)

To this we should add that if others associated the apparatus with the relevant *laws* of electricity, the inventor associated it with his *theory*, which he regarded as "wonderfully confirmed" by his machine.

On the two occasions when he applied this strategy, Volta unquestionably felt—and, in the case of the pile, he successfully convinced others—that the instruments he had built genuinely confirmed the theories with which he explained their functioning; and, more important, that they irrefutably falsified the theories against which he was battling. However, in at least one case—the electrophore—we have reason to believe that Volta realized the limits of his strategy. Yet he did not abandon it. To preserve his instruments' role as reliable witnesses to the truth of his theories, he preferred to modify the cognitive status of these theories. In this sense, the electrophore is emblematic. It constituted a double "imprinting" for Volta, one for the strategy, one for the epistemology he was virtually forced to adopt in order to overcome the difficulties of such a strategy.

Let us pick up the thread of the story. While Beccaria stuck to his course without giving great heed to the objections, Volta too, with no less conviction and stubbornness, continued to study the Symmerian phenomena, repeating and modifying "Beccaria's experiments in order to confute . . . a foundation of his theory" (VO, 3:138). The sequence of events, which unfolded over a period of more than three years, was the following.

After Beccaria had rehearsed the theory of vindicating electricity in a work entirely devoted to the subject (1769a) and in his last major book, *Elettricismo artificiale* (1772), Volta, in May 1772, wrote a letter on the same topic to Priestley. In so doing, he remained faithful to his habit of addressing himself to the foremost European authorities on electricity—a practice he had started with a letter to Nollet and continued with the *Dissertatio epistolaris* to Beccaria. In his letter to Priestley, Volta pointed out the greater "inertia" exhibited by glass compared with electricity "in expelling its electricity and communicating it to other bodies." Volta saw this as

> the explanation of a few other very considerable peculiarities, for example: the fact that a cylinder of hard wood or stick of sealing wax, when rubbed, although they

will attract a light leaf from a greater distance than a glass stick does, do not subsequently repel the leaf with such speed or energy as the glass; and the fact that when we remove and replace the armatures on a wood or resin plate after the discharge, the phenomena of the electricity that has been called *vindicating* last longer, and its signs fade only rather slowly. (VO, 3:141n)

Although the invention was not yet ripe, this letter marked the prelude to the electrophore. Three years later, on 3 June 1775, Volta returned to the subject of vindicating electricity in a letter to Don Marsilio Landriani. He granted his correspondent that it was inadvisable to multiply theories by "constructing one for every new phenomenon"—a failing they both saw in Beccaria. But Volta asked: "Suppose the new facts presented to me were precisely such as to demolish Beccaria's theory and to establish and confirm my own, which I advanced several years ago in a printed letter addressed to that Turin professor: *De vi attractiva ignis electrici* etc. Should I refrain from even alluding to that? Should I report the plain facts and overlook their self-evident application?" (VO, 3:83).

This time the "decisive facts" were not long in coming. Only a week later, on 10 June 1775, another letter to Priestley announced the christening of the instrument of victory with a truly "impressive name": the "perpetual electrophore." This was not the last apparatus for which Volta would claim eternal effects. Even at the close of the dispute with Galvani, he was to produce a "perpetual" electrical instrument, the pile.

The electrophore was a very simple device, consisting of a *dish* and *shield* (fig. 2.2):

> I therefore take a tin plate one foot in diameter, whose edge is raised by just over half a line [1/24 inch]. Inside it I have poured a molten mastic composed of turpentine, resin and wax, spread and hardened into a flat, shiny surface. . . . As an armature above it, I take a piece of gilt wood in the shape of a shield ten inches in diameter (give or take an inch) and about two inches tall. Its base, which must fit the mastic, is flat although somewhat convex on its sides or edge. From the center of the concavity rises a handle of glass or, better yet, of highly polished sealing wax; its angles—and this is quite important—are blunted and rounded. I shall therefore call this armature the *shield*. (VO, 3:95; VOS, 95)

The device was just as simple and "surprising" to operate:

> I charge the plate lightly in the customary manner with the aid of the machine, and cause it to discharge as usual by touching the *shield* and *plate* simultaneously or alternately. I then raise the *shield* by its insulating handle and replace it on the mastic, touching both alternately, as required by the theory of *vindicating electricity*. When I raise [the shield] and when I put it back, the sparks I draw from it are such, and so bright (especially the ones during lifting, and even more so the ones that follow the first two or three) that they spring to my knuckle from an inch and half

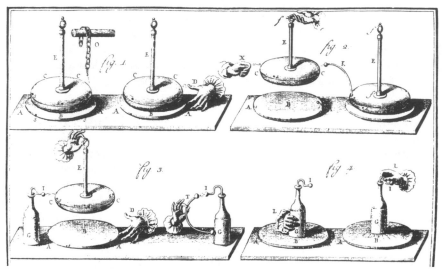

2.2 Experiments with the electrophore. From VO, 3:100.

away and sometimes farther. This is to say nothing of the breeze, and the streaks of light that appear on the points at several inches' distance, and the attraction of tiny bodies over a foot away. (VO, 3:98; VOS, 96)

The conclusion seemed obvious:

> Everything tends to confirm the opinion I already sought to argue convincingly in the Dissertation *De vi attractiva ignis electrici etc.* 1769, to wit, that the electricities of the plates are not actually and completely extinguished by the discharge—as Father Beccaria has claimed, and persists in believing even today—but continue for a long time to adhere to them in part; and that this induces contrary electricities in the respective *armatures*, establishing a degree of balance. (VO, 3:100; VOS, 99)

With that remark Volta terminated his six-year-old dispute with Beccaria. The instrument vindicated the theory. "Our invention of the electrophore," Volta wrote years later, "definitively refuted the system of the above-mentioned author by fully confirming the explanation we gave" (VO, 4:444).

"Explanation" is the significant term here. In his letter to Priestley, Volta argued that the electrophore was a "consequence of the theory" and in "full harmony with the principles"; the previous week, he had told Don Marsilio Landriani that the new facts "established and confirmed" his theory; and just over a month later, in another letter to Landriani, he reiterated the opinion that "my new experiments and discoveries wonderfully confirm those conjectures" (VO, 3:155)—namely, the theory of *De vi attractiva*.

It is easy to see, however, that this interpretation scarcely corresponds to the truth. The theoretical apparatus of *De vi attractiva* was objectively inadequate to explain the functioning of the electrophore—either because of vagueness, as in the case of the saturation concept, or because it was directly disproved, as in the case of the concept of material electrical atmospheres. For that matter, neither in the letter to Priestley—entirely built around facts and effects—nor in his other contemporary letters did Volta offer a *theoretical* explanation of the instrument beyond the mere *experimental* datum that resins possessed less dispersive power.[1] The theory of *De vi attractiva* was mentioned rather than applied; and Volta did not specify the connection between his theory and the properties displayed by the instrument.

Yet this did not stop the instrument from exerting a knock-on effect. Volta not only argued, rightly, that the electrophore sufficed to refute Beccaria's theory of vindicating electricity; he also claimed, wrongly, that the electrophore "wonderfully" confirmed his own theory. Thus he regarded the instrument's practical success as proving the truth of the theoretical construct that went with it. Volta thought so and even wrote so. But it is doubtful whether he remained totally and intimately convinced. The most telling sign of his possible misgivings is his repeated delay in producing the often-promised paper on the electrophore (VO, 3:108, 144). And when—after his decisive encounter with the work of Aepinus and perhaps that of Cavendish (see Gliozzi 1966)—he finally volunteered some fuller indications, the theory was no longer the one alluded to in *De vi attractiva*. Its content and, above all, its epistemological status, had changed. But let us proceed in order.

The first more specific hint at the action of electrical atmospheres is found in an unpublished draft, undated, but probably written in or around 1778:

> To explain, therefore, the phenomena displayed by the Electrophore—which are in fact due to the charging and discharging of the non-conducting layer of resin, and to the exposure of the surface from which the *shield* is lifted—we ought to begin by considering in the simplest manner the action of electrical atmospheres.
>
> Let us take two metal plates *A* and *B* [fig. 2.3]. *A* is insulated and highly electrified *plus*, with two threads hanging from it to serve as an Electrometer. . . . *B too* is insulated, garnished with extremely mobile threads, but not electrified. Let us suspend it (using, for example, thin silk threads, as with the Electrophore shield) from such a height that it will lie outside the sphere of activity of the electrical plate *A*. The suspended plate *B* will give no sign, the attached threads will not diverge nor move toward the finger you bring up to them. And as the suspended plate feels none of the Electricity of the lower plate, so the latter's electricity is in no way diminished or affected by the contacts we make with the former, nor does the divergence of the threads diminish when we do this.
>
> But when you begin to lower plate *A* into *B*'s sphere of activity, you will observe that the supine, parallel threads of *A* will begin to diverge and move toward the finger or any body you present to them; whereas the threads of the electrified plate

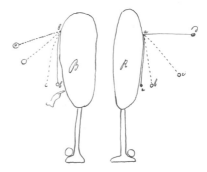

2.3 Volta's conjugate metal plates.
From VO, 1:166.

B will begin to lose their divergence. *B* therefore is affected by *A*'s excess electricity, which it itself takes up—since its already divergent threads flee from another body electrified *plus*, such as a rubbed glass tube or a black ribbon, and rush to a *minus* electric, such as a black ribbon to a sealing-wax cylinder, etc.

One might believe at first that part of *A*'s redundant fire has traveled to body *B*: does that not seem suggested by the lessening of the divergence—which in fact decreases so considerably in *A*'s threads, indeed, by the same extent to which B's threads now diverge? But none of the [typical signs] occurs: the fire does not manifest itself . . . no spark is heard to crackle, nor any ray of light seen to flash.

Yet we could also suppose that the fire flows from *A* to *B* quietly and invisibly.

This assumption, however, is contradicted by the facts, which show that not even the slightest part of body *A*'s redundant fire has gone into body *B*. On the contrary, the fire accumulates in *A*—since you will see that by taking away *B* and removing it completely from *A*, *A*'s threads will return to the first degree of divergence, while B's will lose the divergence they had acquired.

The accidental divergence that is nevertheless displayed by these threads and the excess electricity that affects plate *B* when brought up to a certain distance from body *A* (which is itself effectively overflowing with fire) can be explained thus: although little or none of the electrical fire accumulated over *A* is transmitted to *B*, the *action* of that fire nevertheless reaches out and extends to *B*. Whatever that action is, and in whatever manner we can regard it—as attraction or repulsion—the electrical fluid seems [or: shows itself to be] endowed with both: with mutual repulsion in the parts of the fluid itself, whence its *expandability*; with attraction or *affinity* toward parts of other bodies, whence its tendency to be distributed among all bodies, so that each of these is identically saturated, and that the bodies in which it is poorly distributed attract each other, etc.; the great Franklin has postulated such forces of attraction and repulsion in the electrical fluid, and Aepinus, in his work *Tentamen*, has built a theory on them, reducing these two elements of attraction and repulsion to calculations and subjecting them to formulae. (VO, 3:165–67)

It will be noted that the style of this draft was new. Some concepts of *De vi attractiva*, such as saturation, were barely mentioned. Others were modified, such as the fundamental notion of electrical atmospheres, now stripped of all lingering connotations of orthodox Franklinism and reduced to mere "spheres of activity." Finally, some concepts were questioned, such as attraction—now flanked, albeit tentatively, by repulsion. Volta now followed Aepinus in interpreting the action of the atmospheres, which he no longer even referred to by the old name of "actuation."

But, despite his new approach, Volta's interpretation was not yet precise. In order to make it so, he still had to introduce concepts that were descriptively more adequate and rigorous. For example, once saturation was discarded, what changed in body B when it was placed inside the sphere of activity of A? What remained constant in A, causing its thread pendulum to return to its previous level when B was removed? Volta arrived at the answers in two fundamental papers of 1778 and 1782 as well as other contemporary writings. Here again, his strategy consisted of building an instrument and using it to support a theory. This time, however, the instrument—the condenser—was genuinely a consequence of the theory.

The first writing, *Osservazioni sulla capacità de' conduttori elettrici*, lay the groundwork by introducing the required physical quantities, their basic relationship, and the manner of measuring them. The second paper, *Del modo di rendere sensibile la più debole elettricità sia naturale, sia artificiale*, also called "the condenser memoir," supplied the precise definitions and the theory of the instrument.

The first concept introduced was *capacity*: "Of two conductors of equal surface, the one with the greater capacity is the one in which that given volume is distributed more in length than in width or thickness" (VO, 3:202; see fig. 2.4). Capacity was measured by the number of revolutions of the machine but was defined as a function of two other concepts:

1. *Tension.* This was "the effort made by each point of the electrified body to shed its electricity and communicate it to other bodies" (VO, 3:286). It was "measured by the sign attained by the electrometer's small pendulum" (VO, 3:204).

2. *The quantity of electricity.* This was the aggregate charge or "the sum of all these forces, of all these partial electricities" (VO, 4:418). It too was measured by the machine's rotations or the number of sparks.

As for the relationship between these concepts, "it doesn't take much to understand that the greater capacity occurs where a given quantity of electricity occurs with lesser intensity or—and this amounts to the same thing—where a greater dose of electricity is required for raising the action to a given degree of intensity; and *vice versa*; in short, capacity and action, or electrical tension, are inversely proportional" (VO, 3:286). In other words, to put it even more briefly, $Q = CT$, that is, the quantity of electricity is equal to capacity multiplied by tension. This is the fundamental law of electrostatics and Volta's main theoretical contribution to electrical science.

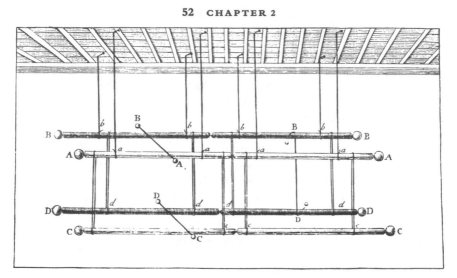

2.4 Parallel-section conductors. From VO, 3:207.

These developments led to the condenser, "an apparatus that enlarges the electrical signs to such an extraordinary degree as to make observable and conspicuous a virtue whose extreme weakness would otherwise cause it to elude our senses" (VO, 3:271). Using the same principles, Volta went on to develop the condensing electrometer (1787)—which not only revealed small charges but made it possible to measure them—and the condensing electroscope (1799), which performed the same function (fig. 2.5).

At this point, Volta's strategy had truly succeeded. What he could not state for the electrophore, he could claim quite justly for the condenser—namely, that "the various experiments that can be conducted with it also shed much light on electrical theory" (VO, 3:285; see also p. 261). Here, unquestionably, the instrument spoke for the theory. And the theory's contribution to the advancement of electrical science was utterly decisive. Never before had the elementary units of electrostatics been clarified with such precision, even if none of the units lacked predecessors. Nowhere else had the fundamental law of electrostatics yet been formulated. But this success had its price. If the instrument shed light on the theory, it was because the theory, with respect to the one expounded in *De vi attractiva*, had undergone a change of content and—more important—of status.

As regards content, we can observe that what emerged from the "condenser memoir" was a different theory. The ideas were either new or revamped. The concept of natural saturation was replaced by that of tension, while even the Franklinian residue of the electrical-atmosphere concept was replaced by fields of forces acting at a distance.

As for the status of the new theory, it was diametrically different from that of

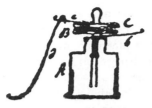

2.5 The condensing electroscope.
From VE, 3:440.

the earlier one. The new concepts were descriptive, not explanatory; thus the new theory was not of the *structural-explanatory* type but of the *phenomenological-descriptive* kind. It was a "black box" theory that stated *how* atmospheres acted but not *why* they did so. Accordingly, the instrument (the condenser) spoke for the theory, because the theory confined itself to a phenomenological description of the instrument without seeking to describe the causes and the internal, hidden mechanism. Volta may have derived this refusal from Aepinus, but he shared it with others. The generation of electricians after Nollet, Franklin, and Beccaria were saddled with the considerable difficulty of explaining the causes of electrical signs with a suitable model. Understandably, this made them lose the confident hope—still voiced by Priestley in his *History*—of finding the true and definitive theory, if only by approximation. Instead, they took refuge under the motto *hypotheses non fingo*. Thus the change of theory was accompanied by a change of epistemology. In Volta's case, structuralism turned into positivism. It is significant, for example, that while maintaining the attraction principle introduced in *De vi attractiva*, he basically lost interest in the question of the nature of forces (were they only attractive, or also repulsive?). Instead, he began to use a language of doubt or compromise (VO, 3:67 and 236n)—which he abandoned in turn. His terse methodological pronouncements therefore became an invitation not only to respect experiments, but also not to go too far beyond them: "Do not engage in disputes and system-building but go to the spot" (VE, 1:484); "do not multiply entities without justification" (VO, 5:82); "reject ideas that transcend experiment" (VE, 5:192); avoid "conjectures that do not tally with experiment" (VO, 6:274). In other words, in positivistic terms, *factis standum*. As we shall see, even Volta's theories of contact electricity, advanced to counter the animal-electricity theory, were consistent with this view.

2.3 NERVOUS FLUID AND ELECTRICAL FLUID

When Galvani began to experiment with frogs—presumably in the latter half of the 1770s—he was certainly not the first to perform such experiments and to observe the contractions of animal muscle under mechanical stimulus or electrical discharge. And when—perhaps in early 1780—he publicly voiced the hypothesis of the identity between nervous fluid and electrical fluid, he was not the first to make the suggestion.

The phenomenon of muscular contractions induced by electrical discharges became well known after the discovery of the Leyden jar and formed the basis of all electrical therapies (see § 1.2). In the hands of physicians, the phenomenon produced no appreciable results so long as electrotherapy was reduced to a blind, uncritical practice. What was lacking was a connection between the therapy and a physiological study—even more important than a pathological study—of the electrical fluid (in making this point earlier, we drew attention to Galvani's greater insight and Sguario's prescience). For lack of interest, but also because of inherent difficulties, the connection was never made. The physiological study, however, did come fully into its own, merging with an older research tradition. Electrically induced muscular contractions soon came to be studied not only in humans, for therapeutic purposes, but also in animals, out of physiological interest.

Sometimes the shift from the practical to the theoretical aspect of the contraction phenomenon was effected by the physicians themselves, with assistance or encouragement from physiologists. Here is a typical borderline experiment, carried out on a human subject and reported in 1759 by Marc'Antonio Caldani, possibly the first to use this kind of stimulus (see Mazzolini and Ongaro 1980, 25):

> In the house of the above-praised Signor Veratti, and on the advice of a learned Professor, an electrification was being carried out on a Man who two years previously had been struck by an apoplexy that had completely paralyzed him on his right side. Signor Veratti kindly wished me to attend this electrification, from which he expected no relief for the invalid, owing to the excessive length of his illness. In fact, after a few days of electrification, the patient grew tired of it, as Signor Veratti had predicted. What was constantly observed, and had already been observed in similar cases by the very learned Signor Jallabert, was that whenever a spark flew from the palsied muscles into an iron held by someone else—at times Signor Veratti, at times myself (because the invalid was overelectrified)—the stimulated muscle contracted sharply, and forcefully drew the parts attached to the back of the muscle toward its front. It was a fine sight to see the mastoid rotate the head, the scalenes bend the neck; the shoulderblade elevator raise the blade; the biceps bend the elbow; in short, to see the force and vitality of all the motions occurring in every paralyzed muscle exposed to the stimulus. (1759, 21)

Analogous contraction experiments, with or without electricity, were performed on animals—usually dogs, cats, and hens—by many other Italian electricians and physiologists. One of the earliest and most frequently cited—among other reasons, for the scientific authority of the experimenter—was Beccaria's, reported in 1753:

> Secondly, it seems certain that the electrical vapor penetrates into the nervous and muscular parts of the animal body, dilating and shortening them; and that its effects are proportional to its density and quantity. . . . But to examine this more

closely I arranged for one of the extensile muscles of a live rooster's leg to be separated from the thigh, leaving it attached to the leg by tendons and nerves alone. I fastened a brass wire to each tendon. I held one wire to the lower coating of the Franklin square; then, after charging the square, I brought the other wire to the upper coating. The body [i.e., the electricity], traveling through a wire into one end of the muscle, and thence down the muscle into the other wire, produced the following effects: (1) At the precise moment the spark burst from the square, the rooster greatly distended its leg, with considerable force; (2) at the same moment, the lateral parts of the muscle body spread outward, dilated sharply and swelled; they began moving toward the tendons, resembling a lady's fan being opened briskly; (3) pricking the same muscle in various spots with a pin never produced a motion comparable to the one just described. (1753, 128–29)

Like Caldani in the human experiment, Beccaria cited Jallabert. But the difference between these experiments and those of the Genevan doctor—which were clearly therapeutic in intent—is obvious from Beccaria's concluding query: "Now, while displaying before our eyes the workings of the electrical body [i.e., electricity] on muscles, might not this experiment add a degree of probability to Newton's queries, principally to his statements on animal motion in query 14?"[2] The intent of Beccaria, and of all who performed similar experiments on animals, was therefore purely theoretical and not medical. The point was to decide whether the electrical fluid could be regarded as an agent of muscular contractions and therefore to understand the mechanism responsible for muscular motion.

This last question, widely debated since antiquity, had given rise to various interpretations. If we refer to the period of Galvani's scientific training and early work, we can distinguish three main theories: (1) the oldest, the theory of nervous fluid; (2) the theory of vibrations of a material medium, which had many variants; (3) the most recent theory, that of the nerveo-electrical fluid.[3] By examining these theories and the accompanying discussions, we will gain an insight into the state of the knowledge available to Galvani. This will enable us to assess more accurately the innovative nature of his own contributions. It should be noted at the outset that none of the three theories was strong enough to establish a secure predominance and—as we would say today—to constitute a paradigm. In 1787, George Fordyce defined them as "mere chimeras of the brain" (1788, 26). The verdict is somewhat harsh, presumably because Fordyce wished to accredit a fourth theory: his own. But all three theories unquestionably contained flaws, some of them very serious. As we shall see, however, a new fact emerged in favor of the electrical theory, although this did not suffice to ensure its unchallenged supremacy.

Let us therefore examine each theory in turn, beginning with the theory of the nervous fluid or "animal spirits." "All of antiquity," wrote Albrecht von Haller in his monumental *Elementa physiologiae*, "has identified in the nerves a very subtle humor or—because the name of 'humor' has an obsolete connotation—a very tenuous fluid, which has been termed *spirit* since it exerts a power that is invisible but as great as air. This doctrine, accepted in the schools, has remained intact over many centuries" (1766, 365–66).

According to this theory, the nerves are extremely subtle canals that convey a special fluid from the brain to the body's peripheral muscles for the purpose of generating motion, and from the periphery to the brain for the purpose of carrying back sensations. Hermannus Boerhaave had established the properties of the fluid using a regressive reasoning that took as its starting point the very phenomena it should have explained. The fluid, he argued, was supposed to be "very fluid, very subtle, very swift and such as to drive its way into the muscle" (1743, 218). Haller too—for the similar purpose of "setting the conditions required for that element to serve as a sentient and motive fluid" (1766, 371)—employed the same regressive method and reached an identical albeit fuller conclusion. The fluid had to be (1) "highly mobile"; (2) "capable of being set in motion by the power of the soul, by the will and by an impression of the senses"; (3) "very fluid in order to allow a very rapid motion"; (4) "so subtle as to flow in tiny canals invisible under the microscope; [so subtle] that the spirits themselves cannot be made visible with any aid from our senses"; (5) "adherent to the nerve so as to remain joined to it and not leave it until its own task was performed"; (6) "unpossessed of taste, smell, color, heat or any other quality that forcibly strikes our senses" (excerpted from pp. 371–73).

Although widely accepted on Haller's authority, the theory of nervous fluid never enjoyed an unquestioned, unanimous support. Haller himself listed eighteen objections, which he examined one by one before refuting them en masse (pp. 366–71). Other scientists presented a smaller number, but regarded them as irrefutable. N. Bertrand, for example, cited five basic objections (1756, 154; see Hoff 1936, 165); and Marc'Antonio Caldani, after discussing six, skeptically concluded that in the face of such difficulties "we should openly admit our ignorance, and not expect human reason to enable us to discover how and with what means the will acts on the brain and nerves (whatever the hypothesis or even the demonstration concerning the structure of the brain and nerves)" (1786, 132).

Other researchers were just as critical or skeptical, perhaps even more so. Robert Whytt wrote in 1751: "But, waiving the objection, that as the nature of the animal or vital spirits, as they are called, is altogether beyond our ken, every account of muscular motion from a *stimulus* which depends on their particular energy or manner of action, must therefore be merely hypothetical and precarious at best" (quoted in Fulton 1926, 33). Bryan Robinson was even more peremptory:

> It has been a received Opinion, that the Nerves are small Pipes which contain a Fluid, called *Animal Spirits*, drawn off from the Blood in the Brain. But it does not appear from any experiments, that the Nerves are Pipes; or that such a Fluid as they conceive *Animal Spirits* to be, is separated from the Blood in the Brain; and there-

fore these Opinions are without any just Foundation. The Nerves are not only impervious to the smallest *Stylus*, but when viewed with a Microscope, evidently appear to have no Cavity. And when we consider the Manner, in which the Favourers of this Opinion have explained *Muscular Motion* by *Animal Spirits*; we must allow, that such a Fluid is altogether unfit for this Work. (1734, 39–40)

While similar objections did not shake Haller's faith in the existence of a distinctive, unobservable nervous fluid, it is understandable that they should have prompted the search for other alternatives. One of these, the theory of vibrations in a material medium, had actually been approved by Newton, who, in query 24 of the *Opticks*, had asked:

Is not Animal motion performed by the vibration of this medium, excited in the brain by the power of the will, and propagated from thence through the solid, pellucid and uniform Capillamenta of the nerves into the muscles, for contracting and dilating them? I suppose that the Capillamenta of the nerves are each of them Solid and Uniform, that the vibrating motion of the aethereal medium may be propagated along them from one end to the other uniformly, and without interruption: for obstructions in the nerves create palsies. And that they may be sufficiently uniform, I suppose them to be pellucid when viewed singly; though the reflexions in their cylindrical surfaces may make the whole nerve, composed of many Capillamenta, appear Opake and White. For Opacity arises from Reflecting surfaces, such as may disturb and interrupt the motions of this medium. (1952, 353–54)

Subscribers to this theory included Robinson[4] and—in a far more cautious and doubting manner—even Beccaria. But it ran into fairly cogent objections. Haller pointed out, for example, that the nerve lacked elasticity, irritability, and oscillatory capability (1766, 361–65). Similarly, Robert Whytt criticized the attribution of elasticity to nervous fibers and judged the oscillations of the ether to be incompatible with observable muscular motions.

Difficulties such as these led most scientists to cling to the traditional theory of the fluid. But this theory itself generated another problem, for it left two main questions unanswered: (1) How does the nervous fluid act? (2) What is the nature of the fluid? To the first question, Haller replied with the positivistic expedient of stating that such an answer was not within his power and that in any case the fluid's "laws of motion" were all we needed to know; to the second question, he responded by producing the above-quoted list of six properties required for the nervous fluid. But neither reply was a *positive* solution to the questions. Understandably dissatisfied, some of the scientists more familiar with electricity and its properties decided to take the big step forward.

The third major theory of muscular motion was introduced precisely by those who, in Volta's words, "were determined to assume that the animal spirits possessed not only the character and nature of an aethereal fluid—whatever it was— but also those specific to the electrical fluid; they therefore eventually declared the two to be a single, identical fluid" (VO, 1:22). The logicality of this identification was confirmed by Haller himself: "Once electrical matter was on every-body's lips and seemed to possess the same velocity as animal spirits and the same power to stimulate even ample motions, people started to think it had the same nature as animal spirits" (1766, 378). Beccaria had already set out in greater detail the arguments for this theory:

> The further experiments and discoveries in electricity—of which Newton had seen only the principle—seem to reinforce the great Philosopher's doubts. The speed with which the electrical vapor moves, changes direction, stops and races forth again seems consistent with the speed and changes in animal sensations and motions. The singular ease of its travel—in general, through electrical bodies by communication, and in particular, through the nervous and muscular parts of animals—is consistent with the ease with which the mutations induced in organs by various objects are conveyed to the seat of sentience; it is also consistent with the agility with which other motions correspondingly ensue in the body. And the contractions and dilations caused in the muscles by an electrical spark or electrical shock are arguments, perhaps even decisive ones, for the above-mentioned conjecture. (1753, 126–27)

The double line of argument used here in the "context of discovery" and the "context of pursuit" is clear: if the electrical fluid possesses some of the relevant properties that the nervous fluid should also possess—such as transmission speed, force, and penetration power—then, *by analogy*, there are grounds for *proposing* the hypothesis that the two fluids are identical; on the other hand, if, as various experiments suggest, the electrical fluid stimulates electrical contractions in dead animals, then, *by abduction*, there are additional grounds for *pursuing* and *supporting* such a hypothesis. This seems to be precisely the reasoning adopted by Tommaso Laghi, who regarded it as "not contrary to the proper scientific method." As he put it:

> In a crural nerve leading out from the vertebrae, severed close to them and nearly dried out after fifty minutes, no more stimulus remained, whereas there was still electricity. Indeed, when irritated with the electrical virtue, the nerve restored motion to the leg. The experimenters accordingly concluded that the nerve was a very efficient conductor of electricity. Since, therefore, electricity flows very smoothly in the nerves and restores the muscles' deficient irritability, it seems to me to indicate the action of animal spirits more than anything else: the electrical effluvium is an adequate proxy for it. I certainly believe that it is in no way contrary to the proper scientific method [*non admodum a recta philosophandi ratione*] to conjecture that the electrical matter—diffused throughout the body by the nervous fluid secreted in the cerebral glands—is so constituted as to flow through the nerves to the senses, fostering motion. (1757, 338)[5]

When did this hypothesis arise, and who introduced it? Laghi was surely not the first, for he himself cited Boissier de Sauvages and Hales. Later, Haller cited

Christian August Hausen, Sauvages, and Deshais, while Volta again referred to Sauvages. It is hard to give a definite answer to such questions of precedence, because the theory must have occurred quite naturally to many researchers. However, it is worth noting that, in Italy, the first reasonably systematic presentation was made by yet another physician, Sguario. His contribution, apparently neglected by most historians to this day, was in fact highly original. Sguario submitted his theory as "my conjecture, which will serve until someone comes along who wishes to demonstrate its unlikelihood" (1746, 358). The fire, he contended, resolves the sulfurous parts of the blood into their elements, transmuting them into a "very subtle, extremely elastic fluid, of the same nature as electrical matter, fire and light. These would be the animal spirits. Generated by the cerebral machine, they would issue in a steady flow with an undulating motion from their reservoirs and the nerve roots, to spread throughout the body—that is, wherever those nervous filaments ran" (p. 360). This very subtle fluid would nevertheless be electrical in the sense in which light, heat, and fire are, "that is, only on account of its substance, which is identical to that of electrical matter," and on account of its effects.

Sguario was fairly confident, if not in his "conjecture," which he himself described as merely "plausible," at least in his method of solving the riddle of muscular motion. This consisted in introducing hypotheses and refuting them one by one until "by dint of excluding so many we shall reach the last." This was precisely the "method for truth-seeking by absurdity endorsed even by Geometers" (p. 358).

Unfortunately, though, Sguario's method could not be the proper one; nor could his hypothesis hold much promise. Chief among its drawbacks was the one that Haller expressed as follows:

> All animal matter is of the same type as the bodies that receive the electrical nature by communication, and all animal parts are equally suited to receiving it. Now let us suppose that the sciatic nerve or the muscle are full of electrical matter and excited into motion; that matter will certainly spread everywhere, unimpeded, throughout the surrounding fat and nearby muscles, until a balance is attained. (1766, 380)[6]

In the face of this truly crucial objection, we can understand why many physiologists, including Marc'Antonio Caldani and Fontana, voiced reservations about the theory of nerveo-electrical fluid. Caldani only admitted that "one may properly suspect the class of stimulants of containing no more effective agent than electrical matter."[7] But he added that both "observations" and "reasoning" had led him "rightly to challenge the supposed electricity of animal spirits"—a property still fraught with "great difficulties" (1757b, 464). Fontana too remained consistently skeptical about the theory. First he wrote that "one must not overemphasize the superiority of electricity's irritative force and conclude from this fact that electricity is the cause of the motion it excites." Hence, "at the moment we cannot decide the identity of the electrical matter and animal spirits" (1760, 206). Later he reiterated his reservations (1767; see Brazer 1963), which he maintained even after Galvani's discoveries (see below, §4.1). In his *Treatise on the Venom of the Viper*, Fontana wrote:

> In a word, we are not only ignorant of muscular motion, but we cannot even imagine any way to explain it, and we shall apparently be driven to have recourse to some other principle; that principle, if it be not common electricity, may be something, however, very analogous to it. The electrical gymnotus and torpedo, if they do not render the thing very probable, make it at least possible, and this principle may be believed to follow the most common laws of electricity. It may likewise be more modified in the nerves than in the torpedo or gymnotus. The nerves should be the organs destined to conduct the fluid, and perhaps also to excite it, but where everything yet remains to be done. We must first assure ourselves by certain experiments, whether there is really an electrical principle in the contracting muscles; we must determine the laws that this fluid observes in the human body; and after all it will yet remain to be known what it is that excites this principle, and how it is excited. How many things are left in an uncertain state, to posterity! (1795, 283)

It must be said, however, that even as Fontana voiced his wish for the future generations, contemporary electricians had already made progress. Indeed, they now had a verified finding that dispelled the main objection to the theory of nerveo-electrical fluid.

In the excerpt just quoted, Fontana refers to the gymnotus and torpedo. In the 1770s, notably thanks to the experiments of John Walsh and John Hunter, the electrical properties of these fish had been securely established. In 1776, Henry Cavendish had offered an explanation of how the torpedo could deliver shocks even if immersed in a conducting body like water, and why it did not exhibit the usual electrical signs such as sparks, attraction, and repulsion. From this, he had concluded that "there seems nothing in the phenomena of the torpedo at all incompatible with electricity" (1776, 222–23). Now this same result lent credence to the theory of nerveo-electrical fluid and made the theory "possible," as Fontana conceded, "probable," for other electricians, or even—for the most convinced—"certified by observation."

"Probable" was how Joseph Priestley qualified it. In *Experiments and Observations on Different Kinds of Air* (first published in 1774), he listed along with electric fish a few other significant phenomena: he spoke of the parakeet's feathers, which seemed "endued [sic] with a proper electric virtue" (1775a, 1:276); he cited Beccaria's experiments on muscular contractions produced by the electrical discharge; and he declared that "the proper nourishment of an animal body, from which the source and materials of all muscular motion must be derived, is probably some modification of the phlogiston" (ibid.). Then—

obviously casting away the empiricist restraint that had led him to recommend the inductive method in the *History*[8]—he concluded:

> My conjecture suggested (whether supported or not) by these facts, is, that animals have a power of converting phlogiston, from the state in which they receive it in their nutriment, into that state in which it is called the electrical fluid; that the brain, besides its other proper uses, is the great laboratory and repository for this purpose; that by means of the nerves this great principle, thus exalted, is directed into the muscles, and forces them to act, in the same manner as they are forced into action when the electrical fluid is thrown into them *ab extra*. (Pp. 277–78)

With this hypothesis, Priestley believed he could explain even "the *light* which is said to proceed from some animals, as from cats and wild beasts, when they are in pursuit of their prey in the night" as well as "the light which is said to have proceeded from some human bodies, of a particular temperament, and especially on some extraordinary occasions" (p. 279).

One of the electricians who, instead, held the theory of the nerveo-electrical fluid to be "certified by observation" was the Abbé Bertholon. In discussing electrical medicine, we have already encountered his lengthy treatise *De l'électricité du corps humain* (1780). This work won prizes from several academies, including that of Lyon, which cited it as "commendable for its profound researches, luminous theory, and numerous new ideas" (1780, vi). The volume was filled with statistics and facts: the publisher assured that it contained "the observations and experiments of more than two hundred and fifty French and foreign authors and scientists." From the theoretical standpoint, the book sought to prove two arguments: (1) that there exists an "electricity specific to the human body and to most animals," in other words, an "animal electricity"; (2) that this animal electricity is produced by rubbing idioelectrical (i.e., non-conducting) parts of the body, such as nerves, bones and cartilage, with anelectrical (i.e., conducting) parts such as fluids and muscles.

But in its empirical foundation, Bertholon's treatise—despite its ambitions—offered nothing different or original compared with the works of Priestley and others. Concerning his second argument, Bertholon himself admitted that it was only "probable"; as for the first, the proofs he did produce were already-known experiments whose connection with the animal-electricity theory was in all cases external, vague, or downright tenuous. Apart from the electric fish, such experiments included the Symmerian phenomena of shirts shining in the dark, the manifestations of electricity on people after a walk, the sparks obtained by combing women's hair, stroking a cat's fur and rubbing a parrot's feathers, the colored halos that appear in the field of vision when the corner of the eye is touched, the attractions and repulsions of gold threads and leafs by men dressed in silk, and other similar cases.

Yet Bertholon was widely quoted and utilized. Galvani himself, in part 4 of

De viribus electricitatis, mentioned him three times (GM, 151, 186, 189; GOS, 288, 314, 315; GF, 73, 86, 87; a fourth reference occurs in one of Aldini's notes to the 1792 edition: GM, 207; GOS, 295). These repeated citations—in a text that contains virtually no others—as well as the identity of opinions expressed, have led some scholars to suggest that Bertholon influenced Galvani (see Sirol 1939, 138–45; Rothschuh 1960a, 33–37). Although one cannot rule this out, there is no conclusive proof. The reference to Bertholon is more probably a token tribute by Galvani—once he had made his own independent discovery— to a figure who was perhaps the theory's leading propagandist. The fact remains that despite their identical assertions, the two theories are totally different in their mechanism. A more important argument for excluding such an influence is that none of the proofs produced by Galvani is to be found in Bertholon.

In reality, Bertholon's experiments were no different from the traditional, highly familiar repertory. They provided little or no support for the theory of animal electricity. A first group, while potentially admissible as evidence, was inadequate, such as the researches on electric fish (on this topic, Bertholon quoted Walsh and Hunter but not Cavendish, who was clearly beyond his reach or area of interest). Others were irrelevant, such as the impressions of color produced by touching an eye. A third category was misinterpreted, such as the Symmerian phenomena and the many cases of electrical signs exhibited by hair, fur, and feathers—which, strictly speaking, indicated the electricity *on*, not *of* animals.[9] Moreover, Bertholon failed to cite the experiments performed by physiologists on dead animals. Yet these experiments, according to the arguments of Laghi and the testimony of Volta cited earlier, were regarded as strong evidence for the animal-electricity theory.

There remained one final, decisive difference between Bertholon and Galvani. Bertholon himself, discussing the influence of atmospheric electricity on the human body, vowed not to seek his proofs in "a vain, murky metaphysics, which must be utterly banned from the realm of true science" (1780, 37). But he hardly ever observed this golden rule. On the contrary, he showed a complete lack of methodological caution in drawing conclusions and of critical measure in evaluating premises. The truth was that Bertholon was not as interested in proving theories as in classifying diseases and indicating cures—or, more accurately, he was interested in theory for purely practical purposes.[10] Indeed, as we have seen, this practical concern prevailed over theory to such a degree that the latter was virtually reduced to the role of alibi. As a result, Bertholon's theory remained almost wholly speculative, bereft of serious proof and even of any sense of proof at all. Rather, it belonged to the species of what Volta called "mere hypotheses and vague theories."

In conclusion, the theory of the nerveo-electrical fluid—propounded by Laghi, Priestley, Bertholon, and a few others—was no more than a feebly substantiated conjecture at the time when Galvani began his experiments. True, it was supported indirectly and analogically by various observations, in particular: (1) the muscular contractions in live persons and animals, where electricity proved the best stimulus; (2) the experiments on muscular contractions in dead animals, where electricity was the only effective stimulus; (3) the existence of fish unquestionably endowed with electrical properties; (4) the manifestations of electricity observed on animals and humans. Nevertheless, the theory was shaky. To begin with, the analogy that served to introduce it was weak, since the very existence of a nervous fluid was presumptive—a typical case of a conjectural theory relying on another conjecture. Secondly, the experiments adduced were inadequate: some—the ones cited by Priestley and Bertholon—were either irrelevant or of scant effectiveness; others, such as the ones on electric fish, could be admitted as evidence only at the price of a hazardous generalization from a handful of animals to the entire animal kingdom; a third category—the contraction experiments—was inconclusive because the fact that electricity was a good stimulus, and in some cases the only one, did not prove beyond doubt that the natural agent of the stimulus was electrical. There were two final weaknesses in the theory of the nerveo-electrical fluid: (1) it had not yet refuted all of Haller's objections, especially the one concerning the fluid's dispersion; (2) it failed to solve some of its own problems, in particular the origin of the electrical fluid in the animal body. Fontana was so right to exclaim: "How many things are left in an uncertain state, to posterity!"

2.4 GALVANI'S EARLY EXPERIMENTS

By 1781 posterity was already at work. At the time Fontana addressed it, Luigi Galvani had already begun his researches. Born in Bologna on 9 September 1737, he was then forty-four years old and professor of anatomy at the Institute of Arts and Sciences of his native town. Galvani's wife, Lucia Galeazzi, was the daughter of Domenico Gusmano Galeazzi, physicist and anatomist at the Institute. She was to play a major role in his spiritual and scientific life (see Pupilli 1956; Mesini 1958, 1971).

For how long had Galvani been conducting his research? And for how long had he been entertaining the notion of animal electricity? His diary, the *Giornale*,[11] which contains the first reliably datable evidence on the matter, opens with an experiment of 6 November 1780, but refers to "frogs prepared in the usual manner" (GM, 233)—apparently an allusion to earlier experiments. In itself, however, this phrase is not enough to substantiate such an interpretation, as the "usual manner" could refer to a traditional practice of physiologists rather than an earlier and original technique of Galvani's. This second reading would seem to be corroborated by a nearly identical expression—a "frog prepared in the same manner"—accompanied by a reference to a similar preparation, that is, a severed frog with its crural nerves bared. The words occur in a letter of Marc'Antonio Caldani of 30 October 1756 read before the Bologna Institute of Arts and Sciences on 25 November 1756 (M. A. Caldani 1757a, 332).

We do know, however, that even before the experiment of 6 November 1780 Galvani had undertaken research on the nervous fluid and embraced the theory of animal electricity. In a list of his dissertations read at the Institute (see Pantaleoni 1966, 34–36), we find a *Dissertazione latina sopra l'influsso dell'elettricità nel moto muscolare* of 2 March 1780. This, as Tabarroni has suggested (1971a, 131), may well be the same paper as the *Saggio sulla forza nervea e sua relazione coll'elettricità* dated 25 November 1782. We shall also see that Galvani made statements in support of animal electricity perhaps even before March 1780—probably in February of the same year.

An obstetrician and physiologist, Galvani was certainly interested in the problems of muscular contractions. This is shown by the titles of two papers read by Galvani at the Institute in 1773: *Dissertazione latina sopra l'irritabilità halleriana* and *Dissertazione latina sopra il moto muscolare osservato da lui specialmente nelle rane*. As for his knowledge of the question of nerveo–electrical fluid, he could not have been unaware of the best contemporary literature—ranging from the opinions of Laghi, voiced before the same Institute, to the objections of Haller; from the reservations of Fontana to those of M. A. Caldani, very likely Galvani's teacher. Galvani was presumably acquainted with the major investigations and with the experimental techniques, since the proceedings of the Institute reported the leading publications on the subject.[12] Finally, Galvani's knowledge of electricity was not secondhand, since he had attended the lectures of Galeazzi, whose daughter he married in 1762.[13]

Galvani's mind was therefore prepared as regards both physiology and electrical science. We shall now analyze how this training was reflected in the *Saggio sulla forza nervea*. This essay will introduce us to Galvani's early concepts of animal electricity, his experimental methods, and his attitude as researcher. It was precisely this attitude that constituted Galvani's main innovation with respect to many other champions of the same theory.

The central theme of the *Saggio* is clearly stated at the outset:

> A not inconsiderable number of anatomists have thought either that the electrical fluid constitutes the undefined, highly subtle fluid that is not unreasonably believed to flow in the nerves, or that it is the nervous fluid itself. I therefore felt it would not be useless to perform various experiments with the electrical fluid on those selfsame nerves, in the hope that they would lead to the discovery of the truth or at least shed some light on the darkness still shrouding the phenomena of nerves. (GM, 3; GOS, 123)

Before attempting to solve this problem, Galvani felt the need to define a methodological framework. Nerves have two actions: sensation and muscular motion. The first is "occult," hidden from observation, while the second is "sensible." The investigation must therefore confine itself to the second and overlook the first. In other words, scientific research must be of an experimental kind, tackling problems for which the answers can be checked by public, repeatable observations. This was obviously an excellent start for an investigation that at the time was very often treated in a speculative manner.

The action of the electrical fluid on the nerves is certainly one of the problems that can be dealt with experimentally. To this end, says Galvani, we must take dead animals and isolate their nerves or muscles from the rest of their bodies. The restriction to dead animals is imposed by the methodological principle laid down earlier: this condition reduces the effects of the electrical fluid (the contractions) to observable causes alone—to the exclusion of the inobservable causes that govern sensations. Frogs are well suited to this purpose. They must be "prepared" as follows (fig. 2.6 and fig. 2.7, lower left): "Cut transversally below their upper limbs, skinned and disembowelled . . . only their lower limbs are left joined together, containing just their long crural nerves. These are either left loose and free, or attached to the spinal cord, which is either left intact in its vertebral canal or carefully extracted from it and partly or wholly separated" (GM, 5; GOS, 125).

By applying the conductor of an electrical machine to the frogs thus prepared, and more precisely to the crural nerves and spinal cord, Galvani obtained his first three experimental results: (1) muscular contractions do not occur when there is a simple flow of fluid from the machine to the nerves and muscles; (2) contractions occur when the nerves receive an electrical impulse that "forms either a spark, or brush discharge [fiocco] or small globe" (GM, 9; GOS, 130); (3) the contractions occur even without these signs provided the electrical fluid is "forced to push in some manner" to overcome air resistance.

Galvani obtained confirmation of these findings by comparing two experiments (fig. 2.8). A frog prepared with an imbedded hook was placed in a glass pitcher sealed by a cork plug. First the hook was connected to a metal wire, and the wire to the machine's prime conductor. In this setup, no contractions oc-

2.6 The "frog prepared in the usual manner."
From GM, 6; GOS, 126.

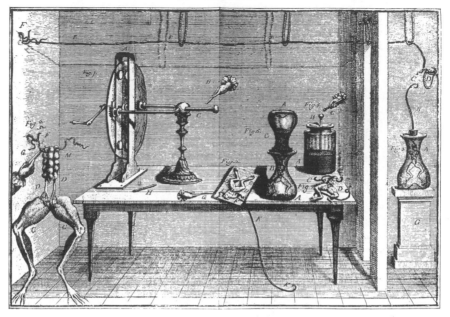

2.7 Plate 1 of Galvani's *Commentarius*.

curred. Then the metal wire was slightly removed from the conductor: in this case, contractions were produced, with or even without an accompanying spark. Galvani explained the difference as follows:

> In the first case, the fluid flows smoothly between the two conductors without encountering any resistance. In the second, it encounters resistance in the thin layer of air between the two conductors; as a result, the electrical fluid is obliged to push in some manner, to make a greater effort to overcome the resistance. Therefore, even an impulse of the electrical fluid equal to the one it generates to make its way through a thin layer of air is sufficient, when applied to the nerves, to excite contractions in their corresponding muscles. (GM, 9–10; GOS, 130–31)

The manner in which Galvani interpreted the two experiments sheds light on the theory of electricity he had in mind. The first experiment was a case of electrification by communication; the second, of electrification by induction. Galvani apparently failed to make a clear distinction between the two, or at least did not seem interested in doing so, since he did not use the conventional terminology either then or subsequently. However, he noted that the contractions required the presence of a dielectric (air) between the conductor and the nerve-muscle system. He also believed that a fluid flow occurred even in the case of induction. There is thus good reason to identify in his interpretation the stamp of a material and mechanical theory of atmospheric electricities, presumably that of the early Beccaria. This theory could not help Galvani solve the main prob-

2.8 The frog prepared in the glass pitcher.
From GM, 10; GOS, 130.

lem of the *Saggio*. Indeed, the essay concluded without mentioning the initial question: Are the electrical and nervous fluids the same? But there is more. Strictly speaking, the *Saggio*'s interpretation of electrical effects in terms of fluid flow leads to a negation of the theory that the two fluids are identical. For this interpretation called forth the same objection that Fontana had raised years earlier against Laghi: If—as Galvani himself observed—contractions required the interposition of a dielectric, how can a nerve carrying electrical fluid contract a muscle in which it is embedded?

Notwithstanding this, by March 1780—if that is the actual date of the *Saggio*—Galvani had already and publicly subscribed to the theory of the nerveo-electric fluid. In the *Commentarius* of 1791 he stated that he had been led to this position only "by reason of our deliberations and numerous experiments" (GM, 166; GOS, 302; GF, 79; see also Aldini 1792a, 235). Now in one of the *Lezioni anatomiche*, presumably dating from Carnival 1780 (see Pantaleoni 1966, 179), we find the following passage:

> We can say that where the research of anatomists ends, that of the chemists begins. Perhaps in only one area do chemists give way to us: in the fact that although they are able with their principles to ascertain the cause of putrefaction, they are unable to establish the cause of the death that precedes putrefaction—a question that is the particular task of anatomists. If they were to see that death occurs when the blood ceases to circulate—and thus to exert the friction on the brain and nerves that produces electrical fluid—and if, like us, they suspected that the cause of death should perhaps be sought not only in the stoppage but in the total extinction of the effect produced by the electrical fluid, who could refuse to be indulgent with them? If their proffered explanation—which they derive from the known, manifest fluid—is not true, it is at least plausible. (Galvani 1966, 137)

A few pages earlier, there is a reference to the nerveo-electrical fluid. At the start of the lesson, Galvani, in the presence of a human corpse, wondered where its altered or putrefied parts had gone, adding: "Lastly, where is that most noble electrical fluid that seemed entrusted with motion, sensations, blood circulation—in short, with life itself?" (p. 135).

Putting together these excerpts from the lecture at the Anatomical Theater and the *Saggio* read to the Institute, we may reasonably conclude that by 1780 Galvani was indeed convinced of the existence of a nerveo-electrical fluid, but had no sound or novel arguments to prove it. At that time, the "deliberations and experiments" later discussed in the *Commentarius* were apparently confined to the customary analogy between the electrical fluid and the supposed nervous fluid, and to the recognition of the electrical fluid's powerful stimulating effect on frog muscles. In this, he did not differ from the contemporaries and predecessors who entertained the same notion. However, compared with at least a few of them—such as Bertholon—Galvani did offer one crucial innovation. His attitude was not speculative: he sought experimental proof. Realizing that the issue was surrounded by "obscurity," Galvani was prudent in advancing his concept of the nerveo-electrical fluid. He regarded it as no more than "plausible" and was willing to debate it (see Aldini 1792a, 236). The very manner in which he expounded his theory or even, as in the *Saggio sulla forza nervea*, kept silent about it was a first sign of his gifts as a critical, cautious researcher holding fast to the canons of the experimental method.

CHAPTER 3

Galvani's Experiments and Theory

3.1 THE "FIRST EXPERIMENT":
CONTRACTIONS AT A DISTANCE AND THE
"RETURNING STROKE"

In his search for proof, Galvani soon stumbled on an important experiment performed "by chance" (Galvani was later to report several serendipities of this kind, sometimes with an exaggeration that has misled his interpreters). The *Giornale* gives the date of the experiment as 26 January 1781—the last in a series begun on the 17th. The previous experiments recorded did not yield any noteworthy findings; in fact, they concerned the direct applications of electricity commonly studied by contemporary physiologists. The experiment of 17 January was, instead, of a special nature and led Galvani nine days later to the discovery of a "new" and "wonderful" phenomenon.

Let us begin with the experiment of the 17th, which consisted in simultaneously electrifying a frog and a Franklin square with a conducting arc connected to the machine. As Galvani described it:

> Exp. no. 1. A frog was prepared with a portion of its spinal cord separated from the vertebral canal, and only its crural nerves connected to its limbs, as shown in fig. A. It was placed on an uncoated glass square. A conductor was run from the square to the spinal cord C, and another conductor to a coated square, fig. 2. Then the machine was rotated. After several revolutions, the conductor running to the spinal cord C was lifted by means of an insulating body, so as not to discharge the coated square. One end of the conducting arc was applied to the lower surface of the coated square, while the other end touched the spinal cord. After letting the machine rotate, we obtained fifteen contractions and distensions of the legs, corresponding to fifteen simultaneous contacts with the lower surface and the spinal cord. (GM, 247–48; see fig. 3.1)

These contractions were explained by the outflow of electrical fluid from the overcharged frog to the undercharged lower surface of the square. From the phenomenon alone, however, it was impossible to determine the identity of the electrical fluid that flowed through the conductor until the balance between the frog and square was restored. Was it the "external, accumulated" fluid that was "held and confined by the frog's nerves and spinal cord"? Or was it a fluid that originated in "the internal part of the nerves aroused by the external part" and

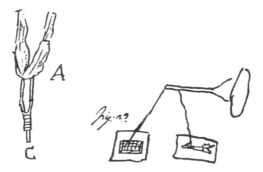

3.1 Electrification of the frog and of the Franklin square. From GM, 248.

"travel[ed] to the lower surface of the armed square, which [was] discharged to a maximum by the charge of the upper surface and by the contact between the conducting bodies" (GM, 248)?

In attempting to solve what had by now become his fundamental problem, Galvani repeated the experiment several times until he happened to observe a new phenomenon: while the frog was lying on the unarmed glass, if someone—"[his] wife or another person"—brought his or her finger up to the machine's conductor and drew a spark, and if someone else—the operator—brought a scalpel up to the frog's nerves or spinal cord, "contractions occurred even though no conductor was connected to the glass on which the frog was resting" (GM, 254).

This is the famous experiment of contractions at a distance, related at the beginning of the *Commentarius* and also known as Galvani's "first experiment" (Fulton 1926, 36; 1940, 302).[1] In the *Giornale* it is recorded as "Exp. no. 6" of 26 January 1781 and described as a "wonderful phenomenon." In 1791 Galvani published his seminal work, *De viribus electricitatis in motu musculari Commentarius*, in the proceedings of the Institute of Arts and Sciences of Bologna, an offprint of which appeared the same year. In 1792 the memoir was published in book form in Modena, with an introductory dissertation and notes by Galvani's nephew, Giovanni Aldini. The *Commentary* describes the experiment of contractions at a distance with a wealth of detail and a lively, incisive style:

> The course of the work has progressed in the following way. I dissected a frog and prepared it as in Fig. Ω, Tab. I. Having in mind other things, I placed the frog on the same table as an electrical machine [fig. 2.7, n. 1], so that the animal was completely separated from and removed at a considerable distance from the machine's conductor. When one of my assistants by chance lightly applied the point of a scalpel to the inner crural nerves, DD, of the frog, suddenly all the muscles of the limbs were seen so to contract that they appeared to have fallen into violent tonic convulsions. Another assistant who was present when we were performing

electrical experiments thought he observed that this phenomenon occurred when a spark was discharged from the conductor of the electrical machine [n. 1, B]. Marvelling at this, he immediately brought the unusual phenomenon to my attention when I was completely engrossed and contemplating other things. Hereupon I became extremely enthusiastic and eager to repeat the experiment so as to clarify the obscure phenomenon and make it known. I myself, therefore, applied the point of the scalpel first to one then to the other crural nerve, while at the same time one of the assistants produced a spark; the phenomenon repeated itself in precisely the same manner as before. Violent contractions were induced in the individual muscles of the limbs and the prepared animal reacted just as though it were seized with tetanus at the very moment when the sparks were discharged.

I was fearful, however, that these movements arose from the contact of the point, which might act as a stimulus, rather than from the spark. Consequently I touched the same nerves again in other frogs with the point in a similar manner, and exerted even greater pressure, but absolutely no movements were seen unless someone produced a spark at the same time. Thus I formed the idea that perhaps in order to produce this phenomenon there were required the simultaneous contact of some body and the emission of a spark. I therefore again applied the edge of the scalpel to the nerves and held it motionless. I did this at one time when sparks were discharged and at another when the electrical machine was completely quiet. The phenomenon occurred, however, only as often as a spark was produced. (GM, 89–90; GOS, 242–43; GF, 45–47)

As Galvani immediately noted in the *Giornale*, the contractions observed in this experiment "scarcely differed from those obtained in earlier ones, where the frog communicated with the machine's conductor or with the square, or received sparks from either the conductor or the square" (GM, 262). It was precisely such a circumstance that suggested that an experiment of this kind could shed some light on the theory of animal electricity. Indeed, Galvani set about repeating it countless times for several years and in the most varied conditions: with the frog close to the machine, far from the machine, in a glass jar, in a glass jar sealed with another jar, in a tin cylinder, in the bell of a pneumatic machine, and so on.

It is almost certain that Galvani, in repeating these experiments, was convinced he was on the right path to explaining animal electricity. Nor is this surprising, if we consider that the unquestionable novelty of the phenomenon must inevitably have exerted a multiplier effect on his cherished expectation, as well as positively strengthening the notion he had already accepted. But the most important question to ask is another one: Did Galvani have *objective* reasons for believing that the phenomenon of contractions at a distance confirmed animal electricity?

To justify a negative answer, it has often been said that Galvani's labors were entirely useless and could have been avoided if only he had been competent in electricity. Volta began by noting that "for one who knows about the action of electrical atmospheres," there was nothing remarkable about Galvani's experiments (VO, 1:46–47 and 175). In his wake, Jean-Baptiste Biot wrote that "reading Galvani's work, one easily notices that he had no knowledge of the true theory of electrical influences" (see Gherardi 1841a, 82). In the same vein, Jean-François-Dominique Arago argued that the phenomenon observed by Galvani "if it had occurred to a skilled physicist familiar with the properties of electrical fluid, would scarcely have attracted his attention" (1854, 212). Earlier, Jean-Louis Alibert, although more sympathetic to Galvani, had written that the phenomenon contained nothing "that ought to surprise an attentive observer, since the explanation is conveniently found in the ordinary laws of electrical influence" (1802a, 38–39; see also Cohen 1953, 27).

The substance of this refrain is that if Galvani had interpreted the phenomenon of contractions at a distance in the light of the best electrical theories then available to physicists, he would not have regarded it as relevant to the issues of animal electricity. Although partly correct, this judgment does not tell the whole story, because it fails to grasp an element that is crucial to understanding Galvani's attitude then and in the later controversy with Volta. Let us take a closer look.

A phenomenon analogous to Galvani's had actually been explained in 1779 by Charles Stanhope (Lord Mahon), leading him to develop his theory of the "returning stroke." When a body, insulated or not (like Galvani's frog), is electrified through the agency of an electrical atmosphere from a charged source (for example, a machine conductor, as in Galvani's experiment), and if the source is removed or suddenly discharged (for example, by drawing off a spark as Galvani's wife had done), the body too will be discharged in the same manner (in Galvani's case, this happened when the grounded experimenter touched the frog with a scalpel). The abrupt restoration of balance causes an immediate reflux, that is, a "returning stroke"—and therefore, in Galvani's frog, nervous stimulation and muscular contraction. A familiar "application" of this theory, identical in principle to Galvani's experiment, was—according to Lord Mahon—the case of the stroke felt by people in the open countryside when the electricity of a cloud is suddenly discharged into the ground through lightning (fig. 3.2).[2]

The obvious question here is: Did Galvani know about this explanation, published just two years before his own observation of contractions at a distance? In the absence of documents, we can glean a fairly reliable indication from Galvani's manner of interpreting the action of electrical atmospheres—the factor advanced by Lord Mahon and the physicists as an explanation of the "returning stroke."

Let us go back now to the *Giornale*. The first observation of the "wonderful" phenomenon dates, as we have seen, from 26 January 1781. There is a silence

3.2 The "returning stroke" in an atmospheric electrical discharge. From Stanhope 1779.

until the 31st, when Galvani recorded two experiments as well as a few initial thoughts and "corollaries." The first experiment mainly consisted in showing that the frog, in the familiar situation, did not contract when touched with glass instead of a conducting arc. In the following experiment, Galvani observed that the frog, even when touched with a conducting arc, did not contract if the nerves were "left with all [their] adjacent parts"; but the spasms occurred when the frog's nerves and spinal cord were covered—except for "a single, small portion"—with a nonconductor such as sulfur. On the strength of these findings, Galvani spelled out six reasons for *ruling out* the action of electrical atmospheres.[3] In particular, he argued that

1. The phenomenon therefore does not seem to be caused by the electrical fluid of the electrical atmosphere of the conductor or disk [because in the conditions indicated the electrical fluid] could neither enter nor accumulate in the muscles or the nerves, and even if had been able to do so through the small portion of bare nerve, it is certain that the effects must have been minimal and barely perceptible.

2. The phenomenon therefore is not caused by the electrical atmosphere of the conductor or disk [because] the phenomenon occurs at the first spark to an almost equal degree as after many; [and] it does not seem certain that such an atmosphere could be expected from a spark, and even if it were expected, the number of contractions should increase in proportion to the atmosphere, that is, in proportion to the revolutions of the machine, but the phenomenon is practically the same at the first spark as at the hundredth.

3. The cause is therefore not the said atmosphere [because] the phenomenon occurs at the first spark even at a distance of six feet, but at such a distance the electrical atmosphere of a spark cannot be transmitted.

4. If it were the atmosphere, this should produce the same and possibly a greater effect even without extraction of the spark, but that does not occur, so it's not the atmosphere.

5. If it were the atmosphere, the phenomenon should occur more easily near the conductor, and after having held the frog a long time under the conductor; but this does not happen; on the contrary, however long I hold the frog under and close to the conductor, I never obtain any phenomenon; but when I place it at a distance from the feet [of the machine], I get the phenomenon as soon as the spark appears, so the cause of the phenomenon is not the said atmosphere.

6. Finally, if the electrical atmosphere were the cause, when a grounded conductor was placed on the glass and when that conductor was in front of and close to the spinal cord, very little or no vapor ought to accumulate; but the phenomenon is the same, the contractions are the same, so it's not the atmosphere. (GM, 256–57)

Having thus ruled out any action of the atmospheres, Galvani drew a few corollaries. The first two contained his explanation of the phenomenon:

> Corollary no. 1. The true cause of this phenomenon is therefore a highly subtle fluid that exists in the nerves, is excited into motion by the impact, vibration and impulse of the spark, and is communicated both to the air and to the highly subtle fluids scattered throughout the air, and to the tiniest parts of the glass or any other body supporting the glass. . . .
>
> Corollary no. 2. This fluid, set in motion and made to vibrate by the spark's impulse and vibrations, is the electrical fluid. . . . (GM, 257)

Turning now to the six reasons that support these corollaries, we can reconstruct Galvani's conception of the electrical atmospheres. These texts confirm the interpretation advanced in §2.4, namely, that Galvani understood such atmospheres as strata of fluid that were released from the charged body and exerted their action through contact or an impact of a mechanical kind. According to this notion, a body immersed in the electrical atmosphere of another reacts because it is directly invested by the excess fluid flowing out of that other body. This is an example of the constructs that, as seen in §1.3, combined Franklin's theory with European theories of effluvia; although abandoned by electricians like Aepinus, Volta, and—in his last years—Beccaria, such theories remained widespread particularly among physiologists, who were influenced by the early Beccaria.[4]

Galvani's notion of electrical atmospheres helps us to understand his attitude toward contractions at a distance. In all likelihood, he must have reasoned thus:

since electrical atmospheres have a direct action and since, in this specific case, that action is either implausible or unthinkable—because if it did exist, the phenomena would be different, as explained by reasons (1) to (6)—then (a) the contractions cannot be due to the fluid released by the machine's conductor; but (b) they must be due to a fluid inherent in the animal, a fluid "stimulated by the extrinsic [fluid] that is applied to the surface."

Neither of these conclusions indicates that Galvani was referring to the "returning stroke": in the negative conclusion (a), he had in mind not the returning stroke, but rather what we might call an *outbound stroke* or, more accurately, a *direct* outbound *stroke* (reason [4] is emblematic in this respect); while in the positive conclusion (b), he was also thinking of an outbound stroke, even if here it was *indirect*—that is, a blow or impact mediated by the air or by the electrical fluid contained in the air.

It must be noted, however, that if Galvani failed to equate the phenomenon of contractions at a distance with those explained by the returning stroke, he did not—in so doing—disregard or violate the available corpus of physical knowledge. The allegation that Galvani was ignorant in electricity is utterly unfounded. On the contrary, he turned to the contemporary science of electricity for a theory that, although hardly the most advanced, was nonetheless shared by many electricians, including a few celebrities: the theory of material electrical atmospheres.[5]

But Galvani was not only familiar with the theory of atmospheres; he intended to apply it and derive physiological consequences from it. On this occasion, he inaugurated a strategy that he consistently tried to implement later on. The strategy consisted in showing that electrical science itself implied, or at least made it possible to assume, the existence of an electricity inherent in the frog. Galvani must have reasoned roughly as follows: if the science is true and therefore, specifically, if the electrical atmospheres exert a material action through fluid contact or transport, then the muscular contractions could not derive from the external fluid's entry into the nerves, but from the nerves' intrinsic fluid under the stimulus of the extrinsic. As for the electrical nature of the internal fluid, Galvani seems to have regarded it as a further corollary of another law of electricity, which postulated that conductors carry fluid. In the case of the frog, the fact that contractions ceased when the nerves were touched with glass or bone reinforced his conviction that the nervous fluid was just as electrical as the fluid that—according to him—sprang from the machine and hit the nerves.

A few years later, however, Galvani modified his views, although it is hard to say exactly when and why. It is fairly probable that Galvani—perhaps with the help of Aldini—had meanwhile perfected his knowledge of electricity. The *Giornale* attests that as late as 1783 he held to his material and effluvial conception of electrical atmospheres. He still regarded the atmosphere, which *in orbem diffunditur* ("is diffused in the sphere") (GM, 378), as a fluid that issued from the conductor, surrounded it, was transferred to neighboring bodies (GM, 379, 382) and acted on these—for example, on frog nerves—through a direct impact (GM, 381). But he subsequently changed his mind.

The earliest evidence of this was in a paper read on 30 April 1789, *De musculorum motu ab electricitate*, in which Galvani explicitly used the theory of the returning stroke explained in terms of electrical atmospheres. Galvani now construed these as areas of electrified air, exactly as defined by Lord Mahon, the later Beccaria, and others. Galvani noted the similarity between the effects of the discharge of a machine conductor after sparks had been drawn and the discharge of a cloud after lightning flashes. He went on:

> So great a resemblance of phenomena and laws seems clearly to indicate that in such a phenomenon the cause and manner of action of the two electricities—artificial electricity and atmospheric, storm-generated electricity—are the same, and in both cases governed by the laws of balance. For in the case of artificial electricity, when the spark flashes, electricity is taken away from the strata of air surrounding the machine conductor; the same seems to occur in the aerial strata surrounding the electrified clouds when the lightning bursts. When this electricity is thus removed, the electricity contained in the bodies contiguous to the strata, owing to its tendency and nature, promptly rushes into those strata, simultaneously restoring their missing equilibrium. (GM, 80; GOS, 216)

Galvani's second reference to the returning stroke came two years later, in the *Commentarius*:

> Now at the emission of a spark, electricity is discharged not only from the strata of air encompassing the conductor of the electrical machine, but also from the nerve-conductors communicating with them. Their resultant electricity is negative. Hence the inner positive electricity of the muscles flows copiously to the nerves through its own strength as well as that derived from external electricity, whether artificial or natural, with the result that, having been taken up by the nerve conductors and having diffused itself through them, it restores failing electricity in them as well as in the strata of air mentioned previously, and establishes itself in equilibrium with the same. This is not unlike in a Leyden jar, when at the emission of a spark, the positive electricity of the inner surface flows copiously to its conductor for the same reason, and then streams out from it, as the appearance of the clear electric brush discharge demonstrates. (GM, 170; GOS, 304–5; GF, 80)

And again:

> As a matter of fact, this phenomenon is accounted for so suitably and clearly by the law of surfaces and equilibrium that I cannot easily dispute with him who also explains those contractions by the same law which are [sic] produced at the discharge of a spark from the conductor of an electrical machine and who considers that there is a double surface, as it were, in the strata of air surrounding this conduc-

tor, an inner one embracing the conductor and an external one embracing the animal. (GM, 171–72; GOS, 305; GF, 80–81)[6]

Thus, a few years years after first observing contractions at a distance, Galvani embraced the then commonly accepted explanation, although he never came around to conceiving the electrical atmospheres as simple "spheres of action," in the manner of Aepinus, Volta, and the later Beccaria.

What was the impact of this change? On the crucial question of animal electricity, none. Galvani never seriously contemplated accepting the conclusion to which the theory of the returning stroke *could*—but did not inevitably—lead: namely, that the fluid whose reflux was thought to trigger the contractions was either the external fluid that had entered the frog, or the frog's own fluid—which, as such, was common to all physical bodies. Indeed, with a keen critical sense, he entertained all the various possibilities. In his paper of 30 April 1789, he recognized that the reflux of electricity might concern "either the electricity contained in the animal, as in a physical body; or the one located in the conducting bodies contiguous to the animal, or the animal's intrinsic one, or the one previously transferred into the animal from the machine or the clouds or from all other sources together" (GM, 80; GOS, 216).

Nevertheless, Galvani held to his idea. Why? Of all the hypotheses on contractions at a distance allowed by the concept of a returning stroke, why did Galvani accept just one—admittedly a notion of a reflux, but a reflux of a fluid specific to and inherent in the frog? It is not enough to say that he had already convinced himself of the existence of this fluid by other means—namely, the analogical-evidentiary approach, which he shared with other physiologists, as we have seen. That statement is true, but not exhaustive. There is another factor, not just psychological, which comes into play here—a factor that already informed Galvani's earlier convictions and was to shape the later controversy with Volta, to the point of serving as a discriminant between the rival positions. This factor is a typical theoretical assumption: Galvani saw the frog's contractions and sought their explanation in an *electrobiological* and not only *electrophysical* context. True, he tried to show that the first was supported by the second. But for him the primacy went to biology and physiology, not to physics or even to chemistry, a science held by Galvani to be incapable of explaining the causes of death and, presumably, even less of explaining vital phenomena (see §2.4). Thus, in Galvani's eyes, the frog was not just an ordinary *physical body*, but a *living organism*. Similarly, applying the well-known abductive reasoning from observable effects, Galvani concluded that the frog *had* to contain not the *common electrical fluid*, disseminated in every physical body, but a *specific electrical fluid*, inherent in and exclusive to the organism.

This is a particularly effective case for illustrating the epistemological notion of the "theory-ladenness" of observation. Galvani's behavior clearly shows the weight and influence of what we can call "interpretative theories";[7] it shows that observation is not the only criterion for determining the field or category in which to place experiments. It was not on the sole basis of what he saw that Galvani opted for the hypothesis of an animal electrical fluid. Rather, it was his biological and physiological outlook, which formed part of his scientific training, that led him to such a conclusion. Indeed, it practically drove him to the conclusion by making it seem so natural to him. All his observations and inductions were informed by this outlook. In our final analysis of Galvani's controversy with Volta, we shall see what it owed to the incompatibility between their different theoretical interpretations about the proper area of attribution of the common, shared observations and phenomena concerning the frog.

Nevertheless, Galvani was not unaffected by the theory of the returning stroke. Interpretative theories, especially for serious, prudent investigators, are not spectacles that allow a view of the world through their tinted lenses alone. Although he did not share them, Galvani admitted the possibility of hypotheses other than the reflux of an intrinsic, specific fluid. Also, he became even more aware that contractions at a distance were not yet a sure proof of the fluid's existence.[8] The result was to increase Galvani's caution.

For that matter, Galvani had always been cautious, even after discovering contractions at a distance. Witness the succession of discordant "reflections" in the *Giornale* over a very brief period of time. The corollaries of 31 January 1781 were repeated in a corollary of 3 February (GM, 260, 261); by 12 February, this had become a "certain corollary" (GM, 272) and remained so five days later, the 17th (GM, 276). But on 31 March another corollary, capping a series of experiments, stated that "the fluid occupant of the nerves is not the electrical fluid; in other words, animal spirits are not electrical fluid" (GM, 287). Later, Galvani reverted to his initial idea (GM, 300).

Yet he did not dispense with precautions. The following excerpt from an anatomical lecture vividly illustrates the combination of deep conviction, uncertainty, and prudence that continued to stir Galvani's mind. The lecture dates possibly from 1786,[9] but is certainly subsequent to the discovery of contractions at a distance, which are explicitly mentioned:

> Nature has openly displayed the magnitude, velocity, and mode of action of this muscular force. But as regards the area to which it belongs, the principle responsible for it, and the laws that govern it, nature seems to have kept silent; or, perhaps, nature has decreed that we should learn these things only after enduring other toils and trials, or by chance or luck.
>
> From the things that have just become clear—although they seem insufficient to us—we can formulate a possibly useful opinion on this question and a hypothesis. This we shall submit to you, young listeners, after observing some of the properties that must be displayed by the principle governing muscular action—whatever the principle turns out to be.
>
> These properties are: extreme subtlety, velocity, and force; also, the power to

transport instantly and voluntarily the particles that compose the muscular fibers; the forced penetration into the nerves; the spontaneous regeneration as new, or recomposition or development or diffusion; the far from insignificant relation with the atmosphere and its vast, sudden changes; the near-instant propagation of motion through the intact nerve and its non-propagation when the nerve is tied or cut; the intermittent action through contact and rubbing of the nerves, which instead becomes a constant action under stimulus; no action, or none of significance, when stronger stimuli are presented to the nerves or muscles; the action excited in the nerves through the mechanical motion produced in adjacent bodies; lastly, the fairly vigorous action on the nerves of a minimal electrical influx and impulse, which can be not only conveyed by a barely perceptible artificial electricity but also induced by a nearly insensible ray of atmospheric electricity and moved by an electrical spark: this impulse, so light and feeble, communicated to the nerves from more than one hundred feet away suffices to excite contraction in the muscles. These and other properties must be inherent in the principle that produces muscular movements, as clearly demonstrated by the new experiments performed and the observation of the phenomena of muscular motion.

After examination of these matters, a conceivably useful opinion may be advanced by someone who voices the hypothesis that the muscular force resides in a nervous fluid, and that this is composed of electrical fluid—not exclusively, however, but combined and mingled with a perhaps very tenuous vapor secreted in the brain and flowing through the nerves. In this manner, the fluid retains some of its properties, loses some others, and produces new ones. And while we submit these opinions, we realize we are talking about things that lack neither obscurity nor great arduousness and that seem little different from those we have already reported at this rostrum.

Among the many difficulties, doubts, and risks of error, only this one seemed promising to us. We have therefore greatly distanced ourselves from the views of some of our learned colleagues and from our earlier opinions, and have chosen this one. We have done so not out of love for discovery or novelty, but to make our earlier proposals more plausible and more useful to our listeners. (GO, 102.4)

It will be observed that in this excerpt the contractions at a distance are not endowed with a special status, but appear in a long list, on the same level—and with the same indicative rather than probative value—as the other standard, well-known phenomena. Although "new" and "wonderful," contractions at a distance did not in themselves constitute a reliable proof of the existence of animal electricity—not even for Galvani. However, they provided a strong indication of it, at least for someone who, like Galvani, had scrutinized them with the eyes of a physiologist. The explanation of the phenomenon afforded by the theory of the returning stroke certainly did not imply such a conclusion, but did not disprove it either. When Volta learned of the phenomenon from Galvani's 1791 *Commentarius*, he immediately offered an explanation in terms of the returning stroke, interpreted in the light of his theory of electrical atmospheres. But even in the *Memoria seconda* (1792), he found nothing incompatible between such an explanation and animal electricity. If anything, the theory of the returning stroke, by demonstrating that different explanations were possible, was an incentive to search for new, more conclusive proofs.

3.2　THE "SECOND EXPERIMENT": THE FROG AS LEYDEN JAR

But how and where could one find other proofs? What direction should one explore? Galvani wrestled with the problem for a full six years without finding any conclusive answers. Meanwhile, he performed the most diverse variants on the experiment of 26 January 1781, which had led him to discover the phenomenon of contractions at a distance. And he might have kept on thinking about new proofs if chance had not put him once again on the right track—this time more efficiently.

Here too, the *Giornale* gives us the starting date for the new cycle of experiments. "Year 1786, 26 April, 20½ hours [8:30 P.M.]—black and white clouds from the south." The meteorological note has its relevance. On that spring evening, Galvani went to the terrace of the Palazzo Zamboni in Bologna with his prepared frog and his usual modest experimental apparatus composed of a few metal wires (fig. 3.3). His purpose was to see if the frog contracted in the same manner when the source was not the artificial electricity of a machine, Leyden jar, or electrophore, but the atmospheric electricity of clouds discharging in a storm.

That is precisely what happened: the experiment was repeated several times, producing the same contractions as those obtained with artificial electricity. But the theoretical structure of the two situations was identical, because the electrical discharge—whether artificial or atmospheric—came from a distant source in both cases. Consequently, the experiment failed to settle the question of whether the agent of the contractions was internal or external. Indeed, Galvani observed that "contractions were obtained during lightning and even when the lightning was not flashing; but by touching the frog, especially its feet, with one's finger, contractions always occurred," even "if one hung a grounded conductor from its feet" (GM, 390). This led Galvani to conclude, as he recorded on 8 June 1786, that "the frog prepared in such a manner is an electrometer for the flow and quantity of the electrical fluid that travels through the bodies contiguous to it—the most sensitive electrometer yet discovered" (GM, 390). An *electrometer*—in other words, a fluid detector, not a fluid *condenser*.

With this experiment, animal electricity actually regressed. It was chance—to use the then customary expression—that decided that animal electricity should take a leap forward instead.[10] A few months later, in early September, Galvani

3.3 Plate 2 of Galvani's *Commentarius.*

stumbled upon a truly new and singular phenomenon. He observed contractions that could surely not be attributed to a discharge of external electricity, there being neither electrical machines nearby nor storm clouds in the sky. This is how Galvani described the phenomenon in *De animali electricitate*, a paper of 30 October 1786—written, therefore, in the heat of the moment:

> In early September, at twilight, we placed . . . the frogs prepared in the usual manner horizontally over the railing. Their spinal cords were pierced by iron hooks,[11] from which they were suspended. The hooks touched the iron bar. And, lo and behold, the frogs began to display spontaneous, irregular, and frequent movements. If the hook was pressed against the iron surface with a finger, the frog, if at rest, became excited—as often as the the hook was pressed in the manner described. (GM, 33; GOS, 163; see fig. 3.4)

Was this not a crucial proof? What other agent, if not an internal electrical fluid, could be responsible for contractions that were entirely similar to electrical ones but were not due to an outside cause? The conclusion seemed obvious, but caution was advisable. Was it really true that the cause was not extrinsic? There were reasons to doubt it, because, since Galvani "had observed these contractions only in the open air and had not yet carried out the experiment elsewhere," he was "on the point of postulating that such contractions result from atmospheric electricity slowly insinuating itself in the animal, accumulating

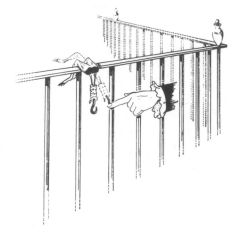

3.4 The experiment with contractions on a terrace railing: the first version of
Galvani's "second experiment." From Sirol 1939.

there, and then being rapidly discharged when the hook comes into contact
with the iron railing. For in experimenting, it is easy to be deceived and to think
we have seen and detected things which we wish to see and detect" (GM, 118;
GOS, 262; GF, 59).[12]

In such a situation, all that remained was to repeat the experiments in order
to isolate the variables better. One therefore had to begin by reproducing the
same situation in the laboratory and then, if possible, perform experiments that
would demonstrate the suspected and desired conclusion in an unambiguous
manner, ruling out all other possibilities. One experiment, carried out a first
time with the hook and a second time without it, proved fundamental. Here is
the version without the hook:

> With one hand, I gripped the spinal-cord canal, holding the frog upright with its
> feet against the upper surface of the silver- or gold-plated box. With my other hand,
> I struck a metal body against that same upper surface of the box or its sides; im-
> mediately the frog began to display the movements that caused it to rise—with the
> result that, when the blows were repeated, the frog seemed to "hop." (GM, 35;
> GOS, 165)

And here is the variant with the hook:

> He [Galvani's assistant Rialpus] grasped the frog's trunk, in which the spinal cord
> had been fastened through the hook, and he kept it suspended from one foot so that
> when the hook touched the upper surface of the silvered box, the foot of the
> opposite leg also touched it. And at the very moment the foot touched the surface,
> all the leg muscles contracted, lifting the leg; immediately after, though, the foot fell
> back onto the surface then rose again; the phenomenon persisted for some time,

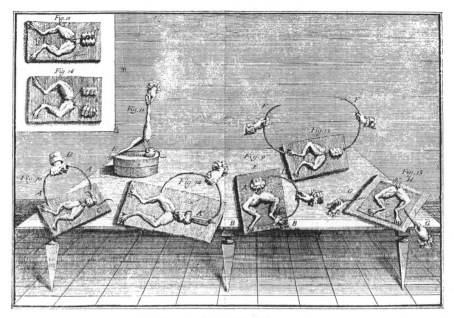

3.5 Plate 3 of Galvani's *Commentarius*.

with the leg alternately dropping to the surface and rising again as if "hopping." (GM, 36; GOS, 166–67; see fig. 3.5, exhibit 11)

This is known as Galvani's "second experiment"; strictly speaking, it consists of two proofs: the frog trunk hopping on two legs and the frog trunk hopping on a single leg—when, in both cases, a metallic circuit is closed. In the *Giornale* these proofs are recorded in the entry for 20 September 1786 respectively as the first and fourteenth of the "experiments with constant contacts and with feet and muscles only" (GM, 400). Both are followed by a virtually identical corollary:

> [*Corollary no. 1*]. The phenomenon therefore seems to depend on a muscular discharge produced by means of the conducting arc, almost as in the Leyden jar. (GM, 401)
> [*Corollary no. 2*]. This discharge seems similar to that of the Leyden jar, when only the [jar's] hook is touched without causing the discharge, establishing a communication between the external and internal surface. (GM, 402)

Thus Galvani saw a structural analogy between a Leyden jar and an innervated frog muscle (fig. 3.6). In both cases there is an imbalance between the external and internal parts that is canceled out when the two parts are made to communicate via a conducting arc. One can therefore say that an innervated muscle is an animal Leyden jar, hence that the animal possesses an intrinsic electricity.

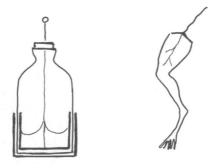

3.6 Structural analogy between frog and Leyden jar, according to Galvani. "Perhaps the hypothesis is not absurd and wholly speculative which compares a muscle fibre to something like a small Leyden jar or to some other similar electrical body charged with a twofold and opposite electricity, and by comparing a nerve in some measure to the conductor of the jar; in this way one likens the whole muscle, as it were, to a large group of Leyden jars" (GM, 153–54; GOS, 291–92; GF, 74).

That was Galvani's reasoning. The double experiment of the hopping trunk made a decisive impression on him. It is no exaggeration to state that the conclusion he drew literally "imprinted" itself in his mind.[13] Henceforth, Galvani would invariably regard the frog not only as an ultrasensitive electrometer but as a Leyden jar, a charged condenser. Nothing ever subsequently distracted him in any serious way from this concept of muscular-contraction phenomena, from this theoretical assumption that became a true gestalt for him—an immediate, spontaneous, natural reading. It is therefore important to examine his argumentation more analytically. Galvani's earlier-quoted paper, *De animali electricitate*, of 30 October 1786 provides us with the material for a logical reconstruction of the crucial steps in the process:

> **1.** *Ascertainment of the initial fact.* The experiment of the contractions observed outdoors on the railing is repeated and shows that when the frog's nerves and muscles are touched by metals and the arc is closed, the contractions occur continuously, albeit with varying intensity depending on the metal.
>
> **2.** *Preliminary hypothesis.* The preceding fact, ascertained for the purpose of controlling the hypothesis that the contractions depend on atmospheric electricity ("however unlikely this seemed") invalidates that hypothesis and confirms or suggests[14] the hypothesis that the contractions are due to "forces in the metals acting on the nerves."
>
> **3.** *First problem.* What type of force or principle causes the contractions?
>
> **4.** *Hypotheses.* This force is electrical (*first hypothesis*); or "some other principle, other than the electrical," is responsible for producing "the frog's motions by means of metals" (*second hypothesis*).
>
> **5.** *Testing the second hypothesis.* The nonconductor/frog/nonconductor circuit is

built in the laboratory. Since, with this circuit, the only force to be impeded is the electrical one, and since no contractions are observed to occur with it, one concludes that these are not due to other forces. The second hypothesis is thus refuted.

6. *Testing the first hypothesis.* The conductor/frog/conductor circuit is built in the laboratory. Since one observes the same contractions as obtained with artificial or atmospheric electricity, and since the circuit does not impede the flow of electrical fluid (unlike the earlier circuit), one concludes that the contractions are an electrical phenomenon. The first hypothesis is thus confirmed.

7. *Second problem.* What is the nature of the force—the electrical principle—that causes the contractions?

8. *Hypotheses.* The force is internal, that is, biological, caused by a natural imbalance between different parts of the animal (*first hypothesis*); or it is external, that is, caused by an imbalance in the metals (*second hypothesis*).

9. *Testing the second hypothesis.* This hypothesis is "utterly inadequate" because it contradicts what has always been "demonstrated by the studies of so many physicists": for all the observations prove that, with the exception of tourmaline, contrary charges do not coexist on a single conductor.

10. *Testing the first hypothesis.* Two experiments are performed:

   a. Experiment of the frog trunk hopping on two legs. The trunk contracts at the closure of a conducting arc represented by the experimenter's body and hands, one touching the spinal cord, the other the metal surface; or by the hands and bodies of a chain of experimenters. The contractions suggest, by analogy with the Leyden jar, that the electrical imbalance lies between different parts of the frog.

   b. Experiment of the frog trunk hopping on one leg.[15] The trunk contracts at the closure of a circuit formed by the muscle of one leg, the hook affixed to the spinal cord and the metal surface. The contractions confirm the existence of an imbalance between nerves and muscles, in other words, that "the electrical fluid, carried by the law of equilibrium from one area to another, forces its way through the nerves, causing the muscles to contract rapidly and the leg to rise."

   The two experiments combined thus confirm the hypothesis of an electricity specific to the animal.

11. *Conclusion.* The electricity that causes the contractions is inherent in the frog. Therefore, animal electricity exists.

Were we to ask if this chain of argument—particularly the key steps (9) and (10)—was stringent, we would already find ourselves in the thick of the controversy between Galvani and Volta. Without yet raising the issue, we should nevertheless highlight a few points that will later prove relevant to the debate.

The first point concerns Galvani's strategy. As already noted, Galvani observed his phenomena with the eye of the physiologist, while seeking to make their interpretations compatible with the contemporary state of knowledge in electrical (electrostatic) science. This makes it even clearer that he was specifically trying to incorporate the laws of electricity into his physiological experiments. In this particular case, his strategy consisted in equating a physiological phenomenon and apparatus with an electrical phenomenon and apparatus. The physiological term of the equation was the frog's contractions when nerve and muscle were connected by a circuit of conductors. The electrical term, well known to eighteenth-century physicists, was the discharge produced when the inner and outer coating of a Leyden jar, or the upper and lower surface of a Franklin square, were connected with a conducting arc. The subsequent course of the strategy was inevitable, for Galvani could argue as follows: *if* physics has properly understood these phenomena and apparatuses, *then* it is *physics itself* that will dictate its laws to physiology and thus require the existence of an animal electricity.

The second point derives from the first but concerns the validity of Galvani's conclusion—the focus of the ensuing controversy. One of the effects of the identification of the frog apparatus and the Leyden jar was to transform the conducting arc into a purely *passive* instrument for revealing and canceling out an electrical imbalance. As a result, those who—like Volta—sought to challenge the existence of animal electricity had to begin by finding physical arguments to revise such an interpretation of the arc. Given the known and accepted laws of conductors (§1.3), however, this would entail a change in one of the most solid parts of contemporary electrical science. Volta's enterprise was therefore anything but easy. Indeed, it seemed quite bold—but not impossible. Galvani's double experiment, precisely because it was subject to the presumption that some of the prevailing laws of electricity were true, did not constitute a *direct* experimental refutation of the hypothesis of an extrinsic balance, that is, of a metallic rather than animal electricity. It therefore left some margin for Volta's attempt.

3.3 ANIMAL ELECTRICITY: ARGUMENTS AND PROOFS

Although intimately convinced of his thesis, Galvani was the first to recognize this situation. Admittedly, the covers of the booklets containing the *Giornale* protocols from 20 September to 20 October 1786 carry the titles "Experiments on the Electricity *of Metals*" and "Experiments on the Electricity of *Metals* in Warm-blooded Animals" (GM, 397, 404, 410; my italics) in his handwriting. However, the paper of 30 October already bore the unequivocal title, *De animali electricitate*, and its conclusion demonstrated that Galvani harbored no serious doubts as to the basic argument.

It is true, of course, that conviction does not constitute proof, as Galvani

knew full well. Not for nothing did he submit the hypothesis of intrinsic electricity to discussion immediately after formulating it, and subjected it to further
scrutiny. His scrupulousness and critical attitude are admirable; his procedure,
which complies with the finest rules of the inductive method, is typical of the
scientist who does not want to be seduced by appearances. To grasp the full
stature of Galvani as a prudent, rigorous investigator, we need only compare his
approach with the adventurous conjectures and hasty conclusions of many of his
"forerunners," including even Bertholon and Priestley.

Galvani's further scrutiny of the animal–electricity hypothesis shows that the
image of the frog as Leyden jar, although bound up with his electrobiological
gestalt, was not an improvisation, less still an uncritical interpretation. Before it
became spontaneous or natural for him, it was built up through a patient, strict
program of experiment and reflection. Our clearest and fullest source for this
program is once again the memoir of 30 October 1786, which practically covers
the third part of the 1791 *Commentarius*. An analysis of the paper shows that
Galvani took into account as many as six hypotheses on the electrogenesis of
muscular contractions, including the animal–electricity hypothesis. The hypotheses covered a fairly broad spectrum, from the most to the least plausible. And
Galvani's sagacity was such that he practically anticipated the main ideas later
adopted by Volta. In summary form, the six hypotheses were:

H_1 Animal electricity with fluid imbalance between nerve and muscle
H_2 Metallic electricity
H_3 Human or atmospheric electricity
H_4 Electricity of nonconductors
H_5 Electricity of dissimilar metals
H_6 Animal electricity with fluid imbalance in the nerve

How did Galvani deal with these hypotheses? The 1786 memoir shows that
he used two different approaches. For some hypotheses, he succeeded in producing full-fledged proofs and therefore a direct experimental test. For others,
he had to settle for plausibility arguments and an indirect, theoretical type of test.
In particular, his refutation of H_2 and H_5 was indirect-theoretical. With his critical acumen, Galvani immediately realized that these hypotheses could explain
contractions just as well and that if he wanted to refute them, he could not
simply oppose his electrobiological assumption or gestalt to the electrophysical
one that informed those hypotheses. He therefore resorted to the best possible
strategy: when he had no direct proof available, he tried to show that the animal-
electricity hypothesis was implied, or at least made credible, by the same electrical laws and theories accepted by physicists. Indeed, nothing seems more
efficient to overcome an objection on principle to a hypothesis than to show that
the hypothesis is fully compatible with that principle. Let us now examine these
proofs and arguments in detail.

Galvani obtained confirmation of H_1 by applying the classic predictive criterion. As he wrote in his paper, "*if* the alleged conjecture of the dual and contrary electricity of nerves and muscles had been true," *then*, by placing the frog on a nonconducting surface and touching its nerves and muscles with a conducting arc, "we should certainly have witnessed movements similar to those we had obtained in other experiments" (GM, 37; GOS, 168). Since the prediction was verified by the experiments, Galvani concluded that "it seems obvious" that H_1 was true.

H_2 could not be similarly treated, because it did not lend itself to special experimental predictions. Galvani therefore attempted to test it on the grounds of external compatibility and intrinsic plausibility. Not only did such a hypothesis clash with the fact—admitted by physicists—that opposite charges could not exist on the same, uninfluenced conductor. It was also "truly difficult" (GM, 37; GOS, 169) to admit that the contractions were due to a current in the arc, because the arc's "extreme subtleness and shortness" prevented it from "exciting such fairly strong contractions" in the frog, "as these usually require a much greater quantity of electricity." Furthermore, even assuming an arc current, it was "even more difficult" to suppose it was replenished with every contraction of the frog. Yet "such a replenishment seems absolutely necessary" because otherwise where did the electricity that produced the "new, more frequent contractions" come from? There was one final reason for implausibility in the metal-electricity hypothesis. An experiment—"Exp. 5" of 13 October 1786, to be precise—showed that "the length of the arc does not seem to have contributed to their [the contractions'] increase, but perhaps diminishes them." "This is worth observing," Galvani noted in the *Giornale*, "because if the electricity came from the metal, just the opposite should happen" (GM, 406).

An implausibility argument, however, is always a double-edged sword. If H_2 was assumed to be implausible, wasn't H_1 as well? Might not an imbalance between nerve and muscle—that is, between two conductors in mutual contact—violate the laws of electricity? The objection was more than relevant; indeed, it was so solid that, as we have seen, Haller and others had long deemed it conclusive. Galvani was well aware of this and therefore had to examine the objection with care. His response was two-pronged: on the physiological level, he introduced special hypotheses to explain the nondispersion of electrical fluid; on the physical level, he cited the existence of other similar cases, while establishing analogies with which he attempted to blunt the objection by turning it from a refutation to a confirmation. In particular, after postponing the physiological part of his reply to the subsequent section, he offered the following four rebuttals:

1. The same unbalanced electricity also exists in other animals (GM, 38; GOS, 169), as shown by "the many experiments performed with them by Physicists" (GM, 59; GOS, 193).

2. Muscles contain a large amount of gummy substance, as does the electrophore plate; hence the frog exhibits the same structure as an electrophore—a nonconductor (muscle = plate) + a conductor (nerve = shield)—so much so that "it would almost be proper to call muscles animal electrophores" (GM, 38; GOS, 169).

3. Nerves and muscles are composed of different substances, unlike metals, which "have the same substance throughout" (GM, 38; GOS, 170).

4. "Our electricity has much in common with that of tourmaline, as regards its seat, distribution, and properties of its parts" (GM, 60; GOS, 194).

Galvani's strategy clearly emerges from these arguments. He replied to the physicists' objections by putting the physicists in contradiction with themselves and then advancing objections of his own. In substance, he reasoned as follows: *if physics is true*—in other words, if, as it claims, electrical imbalance cannot exist on metals, if electricity is produced and conserved only between substances that differ in quality and structure, if there are other animals with contrary electricities, as well as minerals with a double electricity—*if that is true*, then the animal-electricity hypothesis is not only a plausible explanation of muscular contractions. It is also an explanation not contradicted by—indeed consistent with—the accepted theories and laws of electricity.

Galvani applied the same strategy to examining the other hypotheses. H_3 was refuted experimentally in a direct manner, because the contractions occurred even when the arc was insulated with glass, blocking any possible human electricity, or when the frog was immersed in olive oil, blocking out any influence of atmospheric electricity.[16] H_4, on the other hand, was refuted in the same manner as H_2. Strictly speaking, H_4 was untestable, for since conductors can be insulated only with nonconductors, one can never prove that a metal acquires electricity from nonconductors. In any case, the hypothesis was unacceptable on the basis of the same reasoning as before, which ran here as follows: *if physics is true*—that is, if metals do not receive electricity from nonconductors, because this would "introduce something new in physical phenomena"—then H_4 must be false.

The same fate awaited H_5, which later served as the basis for Volta's first theory of contact electricity.[17] Although Galvani had suspected from the very start that the bimetallic arc was "necessary" for contractions, he subsequently assigned it the purely instrumental function of "contrivance."[18] He justified this position with the fact—certified by physics—that "observers had discovered that the dissimilarity of metals had a certain influence in promoting or hindering the electrical explosion in the torpedo." Again, he used the same strategy and type of reasoning. Here, it went: *if physics is true*—that is, if, as it assures us, the bimetallic arc is merely a better stimulus for electrical-fluid flow—then H_5 must be rejected.

There remained H_6. The fact that contractions were obtained even by touching the two ends of the nerve suggested that the imbalance lay in the nerve itself, not between the nerve and muscle. But this hypothesis did not explain why the contractions were also produced by connecting the spinal cord to other fibers *as well*, "that is, tendinous, muscular, nervous, cellular and cutaneous ones." This led Galvani to reject as inconsequential the experiment on which the hypothesis rested. He concluded that one should not attribute to the nerve "any special property of transferring electricity to muscles" (GM, 42; GOS, 174).

Significantly, though, Galvani's reasoning was weak here. Unlike in the previous cases, he did not refer to the findings of physics. He did not state that if physics were true, then H_6 had to be rejected. Rather, he argued that *if the existence of a nerve-muscle imbalance is confirmed by experiment and compatible with electrical science*, then H_6 must be rejected. But the if-clause had no force: there was nothing in a nerve imbalance that could be incompatible with an imbalance between nerve and muscle as well; and in any event the conditionality could have been reversed by giving precedence to the experiment of contractions induced by an arc on the nerve alone. The argument's only possible strength lay in the prior acceptance of the notion that muscular contractions were exclusively attributable to a nerve-muscle imbalance.

This circumstance is enlightening. By the time Galvani came around to examining the nerve-imbalance hypothesis, the image of the frog as Leyden jar had imprinted itself so forcibly in his mind that it emerged as a natural interpretation, impeding and excluding other readings. An imbalance in the nerve contradicted such an image. The experiment that seemed to confirm the imbalance was therefore perceived as irrelevant or, at most, a secondary anomaly. Later, we shall see that anomalies of this very kind, interpreted on the basis of different theoretical assumptions, could give rise to the very readings and conceptions of contraction phenomena that Galvani had rejected.

3.4 "OPINIONS AND CONJECTURES" ON ANIMAL ELECTROPHYSIOLOGY

But let us suppose with Galvani that the truth of the animal-electricity hypothesis had been established on the grounds of electrical science. The battle was not yet won, because that truth had to be vindicated on another ground, that of *electrophysiology*. The "opinions and conjectures" of the last part of the *Commentarius* were designed for that very purpose—for describing in detail the genesis, seat, diffusion mechanisms, and nature of this original electrical fluid in the animal organism. To this end, Galvani used the same strategy previously employed in relation to physicists. In addressing physiologists too, he sought to show—sometimes through proof, sometimes through argument—that his views were not incompatible with their most attested experiments and theories. Galvani's electrophysiological ideas were aptly summed up by Alibert:

1. All animals possess a specific electricity, inherent in their economy, which particularly resides in the nerves, through which it is communicated to the entire body.

2. This electrical fluid is secreted by the brain.

3. The internal substance of the nerves (presumably the thinnest lymph) is endowed with a conducting virtue with respect to this electricity, and facilitates its movement and flow through the nerves. At the same time, the oily coating of these organs prevents the dissipation of the fluid, and permits its gathering.

4. Secondly, Galvani believes that the main reservoirs of animal electricity are the muscles. Each fiber must be regarded as possessing two surfaces, and thus the two electricities—positive and negative. Each of these fibers represents, so to speak, a small Leyden jar, whose conductors are the nerves.

5. The mechanism of all the motions is established as follows: the electrical fluid is drawn out and attracted from the inside of the muscles into the nerves, and then proceeds from these nerves to the outer surface of the muscles, so that each discharge of this miniature muscular electrical jar is matched by a contraction, which is the effect of the stimulus exerted by electricity. (1802a, 80–81; numbering and divisions mine)[19]

Point (1) was an explicit reference to Bertholon, but we have seen the superiority of Galvani's insight and the greater abundance of his proofs as compared with the Abbé's daring ideas. Even on point (2), Galvani had predecessors, as he himself admitted. His views, already expounded in "our public Anatomical Theatre"—that is, in the exercises and lectures in osteology and anatomy of 1780 and 1786—echoed those of Sguario, Priestley, and, above all, Laghi. But he seemed to go beyond his forebears. Indeed, Aldini argued that Galvani's experiments offered grounds for the conjecture that Hallerian irritability itself was the electrical force (GM, 211–14; GOS, 318–19n). Aldini, however, was merely rehearsing an opinion of his uncle already recorded in the *Giornale* entry for 31 January 1781 (GM, 257).

Point (3) has to examined later because it depends logically on (4). As for (4), Galvani solved the problem of the location of contrary charges by adopting the analogy between muscular fiber and the Leyden jar. If, structurally, the frog is an organic Leyden jar, then—he argued—from the known properties of the jar's electricity we can infer, by analogy, an equivalent set of properties for the frog's electricity (fig. 3.7). Now in the Leyden jar the two contrary charges are gathered in the same bottle, and the conductor is the pathway for the fluid from the inner surface to the outer. Thus, in the frog, both electricities must reside in the muscle, which will be charged positively on the inside and negatively on the outside, while the nerve will act as the conducting arc (GM, 152–53; GOS, 291; GF, 73).

As for the fact that the frog was—in structural terms—a true Leyden jar, this could be inferred, again by analogy, from the three different procedures for

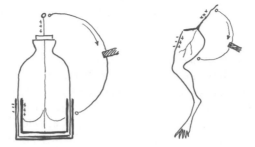

3.7 Structural analogy between frog and Leyden jar, according to Galvani, with charge location and signs: "Now although it is of prime necessity that one apply one end of an arc to the nerves outside of the muscles and the other, as we have indicated, to the muscles themselves, in order to produce muscular contractions, it does not seem to follow, nevertheless, that the nerves are filled with their own electricity, with the concomitant assumption that one kind of electricity has its seat in the nerves and the other in the muscles; just as in the case of the Leyden jar, although we are accustomed to apply one end of an arc to the jar's outer surface and the other to its conductor so that electricity flows from this to the other surface, it cannot be inferred, nevertheless, that the electricity which is present in the conductor is peculiar to that part alone and is dissimilar from that which was collected in the bottom of the jar. Rather it is established that the electricity of the conductor belongs to the inner charged surface of the jar and that both kinds of electricities, although of contrary natures, are contained in the same jar" (GM, 152–53; GOS, 291; GF, 73–74).

discharging the jar. These corresponded to the three ways of obtaining muscular contractions: (a) contact between the conductor and another conducting body (= contact of the armed nerve with a conductor); (b) bringing up the discharging arc (= bringing up the metal arc); (c) discharging the conductor of an electrical machine (= spark).[20]

But however close, the jar-fiber analogy ran into a difficulty, which point (3) of Galvani's theory was specifically designed to overcome. The problem was this:

> Now although this hypothesis and comparison has strong elements of truth, nevertheless there are many considerations which seem forcibly to oppose them. For either the nerves are of an idioelectric (or non-conducting) nature, as many affirm, and for this reason function unsuccessfully as conductors; or they are anelectric (or conducting); and how then can they retain the animal electric fluid within them so that it is not diffused and spread to adjacent parts, with a great diminution of muscular contractions? (GM, 159; GOS, 297; GF, 76)

This was Haller's old objection, which Haller himself, followed by Fontana and Marc'Antonio Caldani, had found so forceful, not to say insuperable. To counter it, Galvani postulated that "the nerves are so constituted that they are hollow internally, or at any rate are composed of some material adapted to carrying electric fluid, and that externally they are oily, or have some other similar substance which prohibits the dissipation and effusion of this electric fluid flowing through the nerves" (GM, 159; GOS, 297; GF, 76). He then sought to defend this hypothesis from the potential accusation of being merely a supporting stratagem. Here too he used argumentation and proofs. He observed that the hypothesis had at least a dual underpinning. The first was theoretical, "since indeed the economy of the animal seems to require that its life force be confined within the nerves"; and this was, with respect to physiologists, a plausibility argument of the same kind as those Galvani addressed to physicists. The second underpinning was observational. "Experiments," Galvani pointed out, reveal that "the nerves are composed of a particularly oily substance," as proved by distillation and the production "of a greater amount of inflammable air from them than was ever derived from any other part of the animal." This air "showed a capacity for emitting a brighter, clearer, and more lasting flame than the inflammable air drawn from other parts." Here was evidence or at least a "clear indication" of "a more abundant oily substance in the nerves" (GM, 160; GOS, 298; GF, 76). At any rate, Galvani added, he would readily reject his hypothesis should "the discoveries of natural philosophers and new experiments undertaken for this purpose bring forward another that is more suitable" (GM, 161; GOS, 298; GF, 77).

In the meantime, whatever the value of the supporting, auxiliary hypotheses, the theory of animal electricity was further confirmed by the fact that this electricity possessed some characteristics "in common with artificial and ordinary electricity and others with the electricity of the torpedo and other animals of this class" (GM, 161; GOS, 298–99; GF, 77). The attested positive analogies of animal electricity with ordinary electricity were six (GM, 163; GOS, 299–300; GF, 77):

1. "An unimpeded and easy path through the same bodies," chiefly metals, and in the same order, from the most conductive to the least conductive; by contrast, "completely blocked paths" through nonconductors

2. "A preference in its emanation for a shorter, easier path, that is, for an arc, angles, and points"

3. "A double and opposite nature, that is, one positive, the other negative"

4. Long adherence to muscles "just as common electricity is accustomed to adhere to bodies by nature"

5. Its renewal, "spontaneous and . . . not restricted to a small interval of time"

6. "An extraordinary increase" in its strength "when the device of arming is employed and when this covering is made of the very metal with which the philosophers are accustomed to encase resin and glass bodies"

There were four attested positive analogies with the electricity of the torpedo and other electric fish (GM, 163–64; GOS, 300; GF, 78):

1. "There is a kind of circuit of electricity from one part of the animal to the other, which is completed either through an arc or through the water itself functioning as an arc, as the natural philosophers have noticed."

2. In both cases, there was an absence of the typical signs of electricity, "the perception of a delicate enveloping atmosphere, as it were, through the attraction and repulsion of very light bodies."

3. Neither electricity provoked "even slight movements in the most modern electroscopes."

4. Neither needed any "preliminary device . . . to excite it, but being prepared, as it were, to manifest itself immediately, it reveals itself by merely a contact."

Naturally, there were also negative analogies, such as the "characteristic and peculiarity of the torpedo in particular, and other related animals, that they can arbitrarily and at their pleasure discharge and eject electricity from their skin in such a way that it completes its circuit outside the body." These facts "indeed indicate that animals of this kind perhaps possess stronger forces and a greater abundance of electricity than other animals, but do not imply that they are, in truth, of a dissimilar nature." Moreover, Galvani added, "perhaps techniques will be found someday whereby effects of this kind may be obtained in other animals as well" (GM, 164; GOS, 301; GF, 78).

Finally, Galvani had to deal with point (5), the mechanism of the fluid's transmission. Here again Galvani was forced to conjecture, because, as he put it, "the manner . . . whereby contractions are induced through the flow of electricity . . . is very difficult to understand and is veiled in obscurity" (GM, 167; GOS, 302; GF, 79). It was not known whether mechanical causes, attractions of particles, or other factors were responsible. By contrast, provided one accepted the analogy between the fiber and the Leyden jar, it was easier to understand how the fluid flowed from the muscle to the nerve and therefore how the contraction occurred.

At any rate, Galvani observed, two conditions were required: (1) "Something that summons the nerveo-electric fluid from the muscle to the nerve and provokes its flow"; (2) "something that absorbs the nerveo-electric fluid in itself as it flows out from the nerve and . . . conveys and restores it, as it were, to the muscles" (GM, 168; GOS, 303; GF, 79).

The first condition is variously met every time the equilibrium of the fluid is altered, for example by touching or rubbing the nerve with a conductor. The second condition is met by any body, such as the conducting arc, that conveys the fluid from the nerves to the muscles. The contraction mechanism must therefore be the following. Touching or rubbing the nerve creates an electrical imbalance. The nerve becomes positively charged, the muscle negatively. When one end of the arc is brought up to the nerve, the arc collects the fluid, and when the other end of the arc is brought up to the muscle, the fluid flows into the muscle. This restores the balance and determines muscular contractions in the obscure manner set out above.

These conjectures marked the close of the *Commentarius* and, with it, of the first step in Galvani's conquest of animal electricity. The only unfinished tasks were (1) to provide explanations for involuntary and pathological muscular movements; (2) to examine the implications of the discovery for pathology and electrical medicine—an aspect we have already discussed. From the first experiment on remote contractions of 26 January 1781 to the publication of Galvani's major work, ten years had passed. His expectations, however, did seem fulfilled and his theory at last firmly established. But just as Galvani was envisaging the applications of animal electricity, others began to challenge those selfsame experimental foundations.

CHAPTER 4

Volta's First Reaction

4.1 "FROM INCREDULITY TO DOUBT"

How was the *Commentarius* received? Let us limit ourselves to early local reactions. All voiced admiration and wonder, but caution prevailed, at least among the most authoritative scientists. The exception was an endorsement from Spallanzani, the probable author of the *Transunto*, a partial annotated Italian translation of the *Commentarius* (S 1792).

Among physiologists, the leading figures who had already studied the question were not greatly disturbed by Galvani's experiments, and tended to reassert their views. Fontana wrote to Marc'Antonio Caldani on 16 May 1792:

> I have read Galvani's work on animal electricity and have repeated most of his experiments, which are true and surprising; there can no longer be any doubt as to the existence of a fluid in animals analogous at least to the known electrical fluid, and the cause of muscular motion in Galvani's experiments. The topic is new, and may prove fairly interesting. However, I find great anomalies that are hard to reconcile with the known laws of common Electricity. This leads me to believe that it is not quite the same principle. (Fontana and M. A. Caldani 1980, 332)

Caldani took an even more openly negative stance, inducing doubt even in his nephew Floriano Caldani (F. Caldani 1792, 166; 1794, 49; 1795):

> Most of Sig. Galvani's experiments succeed, but the first one that prompted him to compose his Memoir is certainly fallacious in kind, number, and case. And it is not very likely that the thing takes place the way he describes it. You seem to have forgotten our work together in Bologna, published in my first letter on irritability and in yours. To put it briefly, by irritating, and in particular by compressing the crural nerves of a frog with any body—insulator or conductor—without any supply [of electricity] from an electrical machine in the room, we obtain violent contractions in the lower limbs. . . . What Sig. Galvani writes in paragraph 6, which begins *At metuens etc.* [GM, 90; GOS, 243; GF, 47], is therefore not true; nor is what he later describes as having deduced from so fallacious an experiment. Without any spark, without an electrical machine nearby, by pricking the said nerves, we produce the contraction of the lower limbs. If the puncture sometimes fails to produce contractions, this is because either the scalpel tip has landed between two fibers, or the conditions for the excitation of irritability are lacking; but the pressure of those nerves invariably, and in the long run, produces the contraction. (Fontana and M. A. Caldani 1980, 335)

Strangely, the response was more positive among physicists. But, at least judging from Volta's case, this favorable reaction was short-lived and more apparent than genuine. True, Volta wrote of his shift "from incredulity to fanaticism" (VO, 1:26). In fact, though, this fanaticism left very few traces. If we consider the actual sequence of his views, it is more likely that the incredulity—which, moreover, went back a long way—was immediately followed by doubt. For this, we need only glance at the chronology.

At an undetermined date, but quite probably in March 1792, Galvani sent the *Commentarius* to Don Bassiano Carminati, who immediately reported its experiments to Volta. From then on, events followed swiftly. On 24 March Volta began his research (VO, 1:3; VE, 3:145); on 1 April he determined the minimum charge needed for the contractions (VE, 3:145) and two days later he corrected Galvani's hypothesis on the location of charges between nerves and muscles, while confirming that "the prepared frog . . . behaves in certain respects like a Leyden jar" (VO, 1:6; COS, 324). Finally, on 5 May, he wrote the *Memoria prima sull'elettricità animale*, in which his enthusiasm had already given way to a modicum of doubt.

Volta's experiments certainly played a role in this change; but perhaps even more decisive was his attitude to the issue. In shifting to his "second manner," Volta adopted the methodological rule of keeping strictly to the facts, without transcending them in bold theoretical constructs (see §2.2). This led him to adopt very severe and prudent criteria for selecting hypotheses and explanations. Such an attitude is borne out by the first part of the *Memoria prima*, where Volta discusses the value of pre-galvanian hypotheses about animal electricity. Although unstated, two methodological criteria clearly emerge from this text. Volta was to remain consistently faithful to them, and we shall see to what extent they conditioned his stance in the controversy—in combination with, and supported by, an interpretative theory of an electrophysical type.

The first criterion regarded the acceptance of hypotheses and may be formulated thus:

> *Acceptance criterion.* The only acceptable hypotheses are those that (a) have direct confirmation; (b) have public, repeatable confirmation.

By this yardstick, Volta held that the pre-galvanian hypotheses about animal electricity should be rejected. First, they violated (a), because, for lack of "true, recognizable signs of electricity," they relied on contrived analogies that were not independently tested (for example, the analogy between nerves and conductors) or were weak (the generalization of experiments on electric fish); furthermore, they violated (b), because the experiments adduced were either "utterly singular, isolated, and unique"—such as that of Domenico Cotugno, who had received a shock from the fur of a live mouse (Vivenzio, in Cavallo 1784, 157)—or repeatable but not with consistent results.

The other criterion that emerges from Volta's discussion concerns the choice of explanations and may be expressed as follows:

Preference (relevance) criterion. If a current, well-confirmed hypothesis explains an experiment, one should not introduce other kinds of hypotheses to explain the same experiment.

Applying this criterion, Volta denied all value to the electrical displays produced on animals and man by rubbing fur, feathers, hair, and clothing. He wrote, for example, that "the rubbing together of feathers, and [of feathers] with the skin, when the Parrot ruffles them, and when they prove to be thoroughly dry, is enough to explain the first fact, *with no need to resort to any specifically animal electricity*" (VO, 1:18; italics mine).

For Volta, this criterion served both to express preference between rival hypotheses and to judge the relevance of experiments. As a preference criterion, it gave the edge to consolidated interpretations, forcing the scientist to reject alternative explanations at least until the current ones had run into insuperable obstacles. As a relevance criterion, its effect was to nullify the probative value of experiments favorable to a new hypothesis when they could be subsumed under an old one. Also, and perhaps most importantly, the criterion imposed directional constraints and therefore had a *heuristic* value. In Volta's interpretation—or, we might say, in the light of his research gestalt—it implied that one should always begin by seeking *physical* explanations, without resorting to other kinds (for example, biological or—a kind he scarcely envisaged—chemical). In other words, it was a conservative criterion, because it resisted any change in the scientific status quo that did not remain within his preferred electrophysical domain. But it was also a prudent criterion, because it opposed the introduction of merely speculative hypotheses and the gratuitous alteration of well-corroborated theories.

Now Galvani's discovery, according to Volta's first reaction, was one that "proves animal electricity with direct experiments and places it among the demonstrated truths" (VO, 1:24). Galvani had proved that the muscular contractions generated by artificial electricity were produced even in its absence simply by applying the arc. These contractions were of an electrical nature, having been obtained with an instrument whose sole function was to correct an electrical imbalance. Thus, in order to explain them, one had to resort to a *non*artificial electricity—an indwelling, animal electricity. Volta wrote:

> Now the conducting arc so simply applied can induce no electricity, as anyone with even a smattering of Electrical Science knows; but its specific and sole function is indeed to remove the existing electricity, to restore balance to the already unbalanced electricity by carrying it from the areas where it prevails in quantity or tension to areas where it is deficient. And precisely for this it is called a *conducting* or *discharging* arc. Must we therefore presume that these organs of the animal happen to be naturally constituted into such a state of electricity—namely, of electrical-fluid imbalance in the relative parts—if the simple conducting arc elicits the above-mentioned contractions in the muscle? I should not say "presume." We must regard

it as certain—to wit, the electrical fluid is the only cause of those muscular motions in such circumstances, and it produces them in no other manner than by finding itself unbalanced between different parts of the animal as described, and by being restored to equilibrium via the conducting arc. (VO, 1:16; VOS, 372)

In this excerpt, and also subsequently, Volta displayed resolve. He stated that Galvani's hypothesis was "proven," was "certain," and was such as "to be no longer in doubt" (VO, 1:27; VOS, 377). But, strictly speaking, Galvani's hypothesis failed to meet Volta's acceptance criterion. For it was proved not directly but *indirectly* on the basis of a presupposition. The substance of the proof, as we have seen and as Volta reconstructed it, was the following: since the conducting arc has only a balancing function, we cannot explain it "otherwise" and cannot invoke "anything else" than an internal electrical fluid. Even the recourse to the "great analogy with the Leyden jar" was just an indirect proof whose value depended on the same presupposition that the arc's function was passive.

It is not at all implausible that, privately, Volta was not enthusiastic about Galvani's discovery, and that he emphasized his support for the sole purpose of expressing later reservations. At any rate, in the light of his preference (relevance) criterion and above all of his interpretation or gestalt, he must surely have felt a spontaneous wariness toward the kind of proof cited by Galvani. Witness the dubitative phrases later in the memoir ("as it seems," "it seems," "as everything leads one to believe"; VO, 1:29, 30 (twice); VOS, 379 (twice), 380); and the promptness with which Volta turned his attention to the conducting arc.

For the time being, though, Volta's experiments and reflections were directed toward three sets of factors: the "quantity, quality and mode" of galvanic electricity. With regard to *quantity*, Volta's aim was to determine the minimum charge required for contractions. He performed experiments using artificial electricity and transposed the results to animal electricity. Volta found that the minimum charge was less than 5/100ths of a degree on his thin-straw electrometer. He also observed that this charge was sufficient to elicit contractions "when, however, the flow of the electrical fluid was directed from the nerves to the muscles, that is, when it enters through the former and travels into the internal substance of the muscles themselves"; by contrast, when the fluid "flows in the opposite direction, that is, travels to the outer surface of the muscle when leaving the nerve, those motions do not occur unless the electrical force is quadrupled" (VO, 1:28; VOS, 378–79).

This finding led Volta to considerations on *quality*, that is, on the location of the charges. He compared two experimental situations, *a* and *b*, relative to the "greater facility of the muscles to contract," and noted that "much less electrical force" was required if—as in fig. 4.1a—"one presents the positive electricity to the nerves that penetrate into [the muscles] and the negative electricity to the outer surface of the muscles, than if we proceed in the opposite manner"—as in

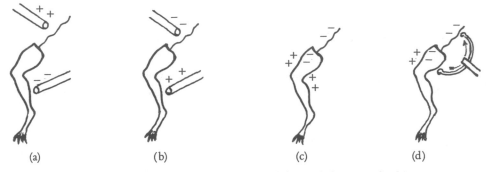

4.1 Volta's experiments on the location of electrical charges in the frog.

fig. 4.1b. From this he concluded that "the organ's own electricity, which causes it to contract through the mere application of the conducting arc, is *negative* on the nerve side, that is, inside the muscle, and *positive* on the outer surface—if, as it seems, this electricity should be regarded as a very feeble charge from a sort of Leyden jar" (VO, 1:29; VOS, 379; see fig. 4.1c). Therefore "the electrical fluid flows from the latter to the former, that is, from the inside to the outside, in a spontaneous or natural discharge of such a kind, and not from the nerve to the muscle, that is, from the inside of the muscle to the outside, as Signor Galvani has claimed" (ibid.; see fig. 4.1d).[1]

But could these findings, obtained from experiments on dead animals, be extended to live animals as well? Volta again introduced expressions that suggested his perplexity:

> Now *if*, as everything leads us to believe, the contractions and voluntary motions of the muscles can take place even in a live, whole animal through the agency of the electrical fluid; and *if*, as we must equally presume, they take place in the easiest manner; then they will occur when the fluid is pushed down from the cerebrum via the nerves toward the muscles. In that case, less force will be required than to draw the fluid up. However, the same movements can also be produced in the latter manner, except that they require a greater force, namely, a swifter or more copious current of electrical fluid. (VO, 1:30; VOS, 380; italics mine)

Volta's doubts were destined to increase when he turned from the quantity and quality to the *mode* of animal electricity. In so doing, he embarked on a series of considerations on the role of the conducting arc that led him within a few days to positions contrary to Galvani's and to those he himself had initially accepted. The starting point was a new experiment. Volta showed that it was possible "through the simple application of suitable armatures . . . to excite . . . in the whole, intact animal the same convulsions, spasms and jerks that are obtained by baring and insulating the nerves, in the manner of Signor Galvani, or using other similar preparations" (VO, 1:32; VOS, 383). The experiment succeeded in the following conditions: (1) when "a piece of thin lead or tin strip" was applied to one part of the frog (spine, back), while "a key, a coin . . . *but of an entirely different* metal from tin or lead" was applied to another part (legs, thighs) (VO, 1:32; VOS, 383; italics mine); (2) when the two armatures were brought into contact "either directly . . . or through a third metal."

The experiment made it clear that, at least in the *whole animal*, the bimetallic nature of the arc was a *necessary condition* and not merely an ancillary, instrumental one. Moreover, Volta was ceasing to regard the arc as a mere indicator of imbalance. This is evident from the ideas suggested to him by the new experiments on the whole animal. The substance of these ideas was a physiological theory of animal electricity partly different from Galvani's. According to Volta's version:

> **1.** The electrical fluid circulates or oscillates "between muscles and nerves and between yet other solid and fluid parts of the body" (VO, 1:33; VOS, 384).
>
> **2.** The fluid is unbalanced but "its motion in the natural state is nonetheless slow, and such that it fails to move the muscles used for voluntary motion . . . and moreover the fluid must cross the intervening animal substances such as muscles, nerves, membranes, and humors, none of which is a sufficiently perfect conductor, and none of which is comparable . . . to metallic conductors" (VO, 1:34; VOS, 385–86).
>
> **3.** No contractions occur "until one disturbs the natural harmonic content of the aforesaid motion of the electrical fluid" (VO, 1:33; VOS, 384).
>
> **4.** This disturbance can be produced by four causes—two internal (act of will, illness), two external (action of artificial electricity and application of armatures connected by an arc).

As a result, the arc now played not only a passive role, but also an *active* one, because, by accelerating the fluid's motion, it became the cause of contractions.

Volta still held to the notion of a weak electrical fluid indwelling in the animal, and did not yet suggest the notion of the arc's electricity. But he now had the proof that Galvani's discovery was falsified in at least one case—whole animals—because his own experiments showed that contractions occurred in them only if a condition that Galvani regarded as unnecessary were met. Thus, there was at least one certain reason for doubt.

In a manuscript probably contemporaneous with the first letter to Antonio Maria Vassalli-Eandi of 1 April 1792 and therefore with Volta's earliest reflections, he not only expressed this doubt clearly, but also outlined an alternative theory: the theory of the electromotive force of metals. Here is how he presented the contending notions:

> All of this is consistent with our ideas about the influence of nerves, and is easily understood. The idea that is not, and for which I have so far failed to find an even roughly satisfactory reason, is the need for *dissimilar armatures*. And, if it is so helpful that one of the armatures should be perfect, that is, extremely well adhesive and, as it were, clinging—why is it not helpful, indeed, why is it actually harmful, for the

other armature to be perfect too? So harmful, indeed, that little or no effect is produced—to wit, that the convulsions can no longer be excited when such perfect armatures are made to communicate with each other? And how is it conceivable that an imperfect one is more effective for this purpose, if this purpose is nothing other than the transfer of electrical fluid from one part of the animal to another? That the two act poorly if both are imperfect is understandable. But that they should also act so poorly, and even worse, when they are both perfect—when each consists of a thin, adhesive metal leaf! This beggars the imagination and is hard to believe. But believe it or not, that is how things are, at least in whole animals, or animals that are merely skinned. (Otherwise, when they are prepared with the nerve left bare and in some manner insulated—as Sig. Galvani does—then the convulsions are excited, as already noted, whatever the type of metal armature applied to the nerve and muscle. So long as the vital forces remain very vigorous, the contractions will occur even without any armatures, by the mere contact of the conducting arc, as we have already remarked.) That is what happens, I repeat, when the nerves are kept covered. In other words, the influence of these similar or dissimilar armatures is such that the effects—the usual movements and convulsions—will be more easily obtained by applying the armatures to similar parts of the animal, that is, by applying both of them to muscles, indeed to companion muscles, such as the gastrocnemius of each leg, and even to two parts of the same muscle, provided the armatures are dissimilar; rather than by applying similar armatures to dissimilar parts of the animal, for example one to the backbone and the other on the leg muscle, despite the fact that these are the best locations, given the correspondence between nerves and muscles, as we saw above.

Now, in reflecting on all this I sometimes begin to wonder whether the metallic conductors, when they are different or are applied differently to two parts of the animal, do nothing other—when a communication is set up between them—than form a pathway for the electrical fluid, which naturally tends to travel from one point to the other, as we are apparently to believe. In a word, I begin to wonder whether they are merely passive agents or instead positive agents, namely, agents that shift the animal's electrical fluid at their own initiative, and, taking it in its quiescent and balanced state, disrupt this equilibrium and cause the fluid to enter in one place through an armature of one kind and exit in the other place through an armature of another kind. I cannot help entertaining such a suspicion when I see the convulsions—and they are strong ones—produced in the frog's legs when, by means of a conducting thread, I establish communication between the tin leaf I have wrapped around part of one thigh and the silver coin placed at the corresponding spot on the other thigh. Here there is no reason why the electrical fluid should tend to flow from one thigh to the other, or vice versa. Indeed, nothing seems to occur if the armatures are similar—two coins, or two leafs. Thus, if convulsions occur, indicating a current of electrical fluid—that is, a transfer from one thigh to the other—when the leaf is applied to one thigh and the coin to the other, we are apparently to infer that all of this is due to the action of the armatures; that the armatures—depending on the differences in the closeness of their fit, in their roughness or smoothness, flexibility, etc.—generate the imbalance of electrical fluid at their points of contact; in short, that the same thing occurs there—albeit to a lesser degree—as occurs through rubbing, which is the means of exciting artificial electricity. For in the latter case, we observe that when a body is rubbed with dissimilar metals, or even a single metal, the metals give some of their fluid to the body or take some from it, depending on the degree of pressure, heat, flexibility, roughness, or smoothness of the metal blade. As a further reflection on the matter, we can observe that it is not even necessary to perform rubbing in the strict sense; if the circumstances are right, a blow, an impact or even a small pressure of any kind will suffice. (VO, 1:39–40)

With these arguments Volta threw Galvani off balance. While Galvani rested his entire theory on physics—in particular on the purely conductive function of metals—Volta hinted at a transformation of physics itself by disclosing the possibility of a new function of metals: the electromotive function. In so doing, he removed the main underpinning of Galvani's strategy. With a different physics, the hypothesis of animal electricity could no longer be so certain and might even have to be withdrawn altogether.

Other doubts might arise from a different register of ideas, in which the pref-
ce criterion again played a part. Volta had found that Galvani's first experi-
—the contractions at a distance—could be explained in terms of *his*
's) electrical atmospheres, construed in a nonmaterial way,[2] without re-
e to any animal-electricity hypothesis. This proved that a phenomenon
dy transferred to the theoretical framework of animal electrical science
uld be maintained within the familiar framework of physical electrical sci-
e. A further implication was that other phenomena—such as Galvani's sec-
periment—might be explained in the same manner.
m up, Volta's first critical reaction to Galvani's discovery seems to have
d in the following sequence:

Volta observed that Galvani's second experiment—however varied, rigorous,
ic, repeatable, and consistent—did not suffice to rule out an active function of
onducting arc. This was one reason for doubting Galvani's explanation.
, Volta further observed that there was at least one case—the contractions of
le animals—in which the bimetallic arc unquestionably performed an active
ction and was not just a passive instrument for correcting an imbalance. This was
n even greater reason for doubting the existence of a purely intrinsic electricity
naffected by external factors.
3. Finally, Volta observed that Galvani's first experiment could be kept within
he theoretical framework of physical electrical science. His preference (relevance)
riterion obliged him to regard these experiments as irrelevant for the purpose of
e animal-electricity theory; it certainly prompted him to adopt the same electro-
hysical framework for the second experiment as well.

The kernel of the emerging conflict can already be defined as a show of force between two research guidelines, assumptions, or gestalten, as we have also called them. Galvani pressed for the inclusion of the contraction phenomena in the field of electrobiology, the only prerequisite being their compatibility with the known laws of electricity. Volta, instead, urged the inclusion of the phenomena in a new science of physical electricity.

This clash also marked the contours of each scientist's research domain and central preoccupations. Galvani would henceforth have to seek to isolate his own domain by showing its peculiar irreducible properties. Volta, instead, would have to seek to enlarge his own to the point of including his adversary's. Thus, one militated for differences, the other for identities; one viewed the frog as a living organism, the other as a generic physical body. As a result, *ceteris paribus*, the controversy shaped into a confrontation between assumptions or interpretative theories—*ceteris paribus*, because experiments came first.

4.2 "IT'S THE DISSIMILARITY OF METALS THAT CAUSES IT"

When Volta resumed work in May to draft the *Memoria seconda sull'elettricità animale* (14 May 1792),[3] the situation appeared to him as follows. The first two parts of Galvani's *Commentarius* seemed "useless" (VO, 1:47n; VOS, 394n); the third appeared at least doubtful in its conclusions; as for the fourth, Volta had already elaborated a partly different electrophysiological theory. While Volta tried to acclaim "the great, the wonderful discovery of Animal electricity innate and specific to the organs" (ibid.), his belief in the discovery seemed to grow shakier with each new tribute. In fact, by May, his faith had practically collapsed, because—even if he concealed it—he had already prepared an alternative explanation of the contractions.

At the outset, however, despite the lack of direct proof, Volta regarded the existence of animal electricity as at least "conceivable." And nearly half the *Memoria seconda* is dedicated to demonstrating this plausibility. In the first place it was suggested by the research on the minimum charge required for contractions. Since the charge turned out to be too small to be detected even by the most sensitive electrometers, Volta concluded that "if so much can be accomplished in the animal's organs by an electricity so weak as to elude all Electrometers, it is no longer difficult to conceive that the same effects—namely contractions and muscular movements—can be produced by an equally weak electricity specific to, and innate in the organs, in other words, an electricity of such weak tension as to be incapable of moving the most delicate of those Electrometers" (VO, 1:55; VOS, 404).

The plausibility of animal electricity was further strengthened by the analogy with light, "which, even if not endowed with a mechanical momentum capable of producing the slightest sensible impulse—of moving, for example, a feather or other very light body invested by it—nevertheless strongly excites the optic nerve, to the point of irritating it with too lively a sensation; indeed, even a feeble and scarce light will excite the nerve to no inconsiderable degree" (VO, 1:56; VOS, 405).

But the experiments proceeded apace, and no sooner did Volta argue that Galvani's discovery was conceivable than he was obliged at least to redimension it. The two following experiments were decisive in this respect:

> I prepare the leg of a big frog so that the crural nerve is properly bared and excised from the backbone and runs out of the thigh along its entire length. I cover one end by folding a small metal strip around it, or I attach small clips to it. I do the same to another piece of the same nerve a little bit below, that is, I wrap it with another thin ring-shaped metal strip or I pinch it with other clips, leaving an interval of one or two lines between each such armature, so that another portion of bare nerve is left below the lower armature, like the small part left bare between the two. I then discharge a very feebly charged Leyden jar—that is, exhibiting few or no sparks—over the two armatures placed on the nerve, so that only the nerve segment lying between them is in the *circuit* of the charge [fig. 4.2a]; and suddenly all the leg muscles are seized by convulsions, and it kicks and jumps—despite the fact that, as is clearly evident, the current of electrical fluid is confined to the nerve alone, indeed to a small part of it, and the muscles and the whole leg have been totally excluded from it. The current, therefore, does not have to reach the muscles in order for the stimulant electrical fluid to invade them; it need only tickle and stimulate the nerves on which such muscles, capable of voluntary motion, depend. . . .
>
> We have now shown what happens when artificial electrical charges are employed. Although in these proofs the electrical fluid restricts its action to just a few points, and to a small section of the nerve, nevertheless it generates contractions and motions of the muscles that—however distant—obey the nerve. The same also happens with discharges or flows of electrical fluid that are not produced by any previous artificial charge but travel from one part of the animal to another simply through the application of suitable armatures and the conducting arc. What happens is that when the action is exercised on the nerves alone, in fact on a small segment of the nerve trunk, there is a response in the motion of the muscles subservient to those nerves, even though the actual electrical current does not reach those muscles. Let us strip and insulate the crural nerve of a frog, the sciatic nerve of a Lamb, etc., and as above . . . let us apply to two fairly close parts of the same nerve the metal armatures, one of tin leaf, the other of brass or, better, silver (we shall soon see how important it is that they should be of different metals); let them be made to communicate either through the agency of a third metal or even without this by moving one toward the other until they touch [fig. 4.2b]: instantly the entire limb will be excited into convulsions and kicks; yet the limb has not been touched, and

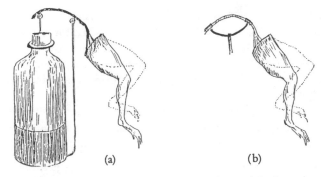

(a)   (b)

4.2 Volta's experiments on the action of the electrical fluid on nerves.

it is inconceivable that it could be reached by the electrical fluid, which has traveled only between parts, indeed, between two adjacent parts of the nerve. (VO, 1:58–60; VOS, 408–10)

In the light of these experiments,[4] Volta abandoned all restraint and decided to contradict Galvani. According to Volta, the experiments proved

1. that the electrical fluid acted on the nerves and not, as Galvani believed, on the muscles, because the muscle lay outside the circuit whether artificial electricity (fig. 4.2a) or the bimetallic arc alone (fig. 4.2b) was used to produce the contraction;

2. that, consequently, "the greater step" taken by some physiologists, notably Galvani—namely, of identifying the animal spirits with the electrical fluid that acts directly on the muscles by flowing through the nerves—was one of those "plausible and seductive" explanations that nevertheless had to be withdrawn in the face of contrary experiments;

3. that, despite this, the muscular fiber could not be equated with a Leyden jar, as Galvani claimed, because the imbalance already existed in the nerve.

One might have objected that the experiments from which these conclusions were drawn were contradicted by another series of experiments—those in which contractions were obtained in the whole animal without baring the nerves, but simply by applying dissimilar armatures to the muscles and placing them in contact. For Volta, however, the objection could be easily countered with the following argument:

But might there be no nervous ramifications in each of those muscles? Might they avoid the electrical puncture, so to speak? I have effectively shown above . . . that when a small electrical current travels through a piece of raw nerve, without muscular fiber, and stimulates the nerve, the latter excites the convulsions and movements of the limb it controls, even though the current does not reach the limb's muscles. But I defy anyone to prove to me that when a current of electrical fluid pervades a muscle, or even just a part of it, it does not strike any of the nervous filaments scattered therein. If this cannot be demonstrated, my proposition remains safe and sound: it is the nerves that are excited by the flowing electrical fluid. (VO, 1:62; VOS, 413)

"But there is more," as Volta immediately added. There was the fact that, while no experiment could prove that the electrical fluid acted directly on the muscles, there were direct experiments proving the contrary.

But there is more: while you produce such experiments as objections against me, arguing for the direct action of the fluid on muscular fibers, you cannot reduce the phenomenon to the point where such direct action is evident and perceptible from the current's striking the fibers alone and provoking the contraction. As a result, there will always remain at least a doubt as to whether a weak electrical current— which is the sort we are dealing with—is of such a size that I can, with direct experiments (which I am saving for discussion for the end), make it perceptible and manifest for you, perceptible to your own organs—yes, let's say it even in passing!—perceptible to your tongue, even to your nerves. For it is the nerves that feel not only the current of electrical fluid that springs forth in the shape of the brush discharge, producing the familiar fresh breeze on the tip of electrified conductors; but also the other, invisible current of the same fluid, which is produced by the mere application of suitable armatures that are then made to communicate with each other. It just takes a simple device: apply a shiny, clean strip of tin or lead to the tip of your tongue, and place a gold or silver coin, a silver spatula or a spoon in the middle of your tongue. Then, touch the handle of the spoon or spatula, or the coin, to the tin or lead blade pressed by the tip of your tongue. This is all it takes, I assure you, to feel the same acidulous taste that you get on your tongue when you place it in front of the tenuous brush discharge and breeze of an artificially electrified conductor at a sufficient distance not to draw sparks. Thus, even here, the flow of electrical fluid across the tongue occasioned by the mere application and contact of two metals excites the very same sensation, the same acid taste—indeed, not weak but lively—but no contraction, no other motion in the tongue, despite its high mobility and irritability. This is quite sufficient to prove that it is the tongue's nerve buds, not its muscular fibers, that are immediately affected in either case by the electrical fluid, which excites and stimulates them by a gentle inflow. (VO, 1:62; VOS, 413–14)

In June, Volta performed a slightly different version of the experiment. To dispel the suspicion that "what we taste is the metal's own flavor licked by the tongue" (VE, 3:171), he dipped the tip of his tongue in a glass of water containing a tin blade. When a silver spoon formed an arc between the middle of the tongue and the tin blade, the same acid taste was felt. This proved that "it is not . . . the metal's flavor that is tasted by the tongue dipped in water, but the flavor of the fluid that flows through the metal and water, namely, of the electrical fluid" (VE, 3:172).

These experiments raised a problem and, at the same time, suggested a possible solution. The problem was the following: "It is also hard to conceive how the electrical fluid moves between two points of the same nerve that are so close together, simply through the application of the armatures and the establishment of an external communication between them; and why this requires *dissimilar* armatures. But this fact is proved by direct experiments, and will be discussed elsewhere" (VO, 1:60; VOS, 410). As for the possible solution, it could only be the one derived from the experiments themselves, namely, "the action of the *dissimilar armatures*, which is not yet properly understood, but has been factually ascertained with all manner of proof" (VO, 1:60; VOS, 411). And indeed the experiments proved that (1) the phenomenon was electrical, at least if one reasoned on the same effects/same causes basis, because it exhibited the same signs as artificial electricity (muscular contractions of the frog's leg, acid taste on the tongue); (2) such an electrical phenomenon manifested itself when the armatures were dissimilar; (3) therefore, "it's the dissimilarity of metals that causes it," because if the agent had been simply an unbalanced animal electricity, then the single-metal arc would have been enough to reveal it.

Two points are worth noting here. The first concerns Volta's attitude. Volta was still unable to refute Galvani's discovery completely. Indeed, he could not have done so, because his experiments did not prove that dissimilar armatures were *always* needed for contractions—since there were cases where even the single-metal arc sufficed. All Volta could do, therefore, was to redimension and circumscribe the discovery of animal electricity. In other words, his experiments allowed him to raise a reasonable doubt but not yet to generalize it with certainty. This may explain his persistent caution even in the *Memoria seconda* and his reluctance to commit himself in public to the general hypothesis of the electromotive power of metals already formulated privately, as in the manuscript quoted in §4.1. In the event, Volta's reluctance was short-lived, and he very soon came out into the open.

The second point relates to Volta's method. Volta arrived at his hypothesis from two different starting points: (1) Galvani's own experiments, interpreted in a manner not entirely consistent with Galvani's theory; (2) new experiments.

(1) Volta began with the experiment in which contractions were produced using the arc on the nerve alone. Galvani himself had performed the experiment back in 1786 using only the single-metal arc. However, he had refused to conclude that the imbalance resided in the nerve alone. The official reason he gave was that the experiment failed to explain the presence of contractions "also" when the arc was applied between nerves and muscles. But the main reason, as we have seen, is that the experiment clashed with his predetermined concept of the muscular fiber as Leyden jar.[5] Moreover, Volta used *dissimilar* armatures in his experiments. Galvani too had conducted experiments on dissimilar armatures and bimetallic arcs, often observing their effectiveness as "contrivances for exciting muscular contraction" (GM, 61; GOS, 196).

(2) Volta performed an experiment in which he produced muscular contractions with whole animals and dissimilar armatures. He also succeeded in stimulating the "sensory nerves" (first the tongue, then the optic bulb) with artificial electricity and dissimilar arcs. These experiments were new, although the second not completely so, since—unbeknown to Volta—it had been carried out years before by Johann Georg Sulzer.[6]

Volta's crucial argument for redimensioning Galvani's discovery hinged precisely on experiments (2). The argument must have run roughly as follows. If, at least in certain cases, dissimilar armatures are not only useful but necessary to obtain contractions (the sure sign of electrical-fluid flow), then one cannot say that such a fluid flow is caused by a *natural* imbalance existing in the animal, for this would also be revealed by a single-metal arc. Rather, the flow must be caused, in a manner still to be discovered, by an *artificial* imbalance created by the armatures. In other words, if the unlikeness of the armatures is a necessary and not just favorable condition for contractions—as shown by the fact that when the first are missing, the second do not occur—then the electrical-fluid motion depends on the artificial armatures rather than on natural conditions. As Volta later wrote: "When such a circumstance is necessary—to wit, if the armatures must be dissimilar in order to excite convulsions and movements, so that when the armatures are identical the motions no longer occur—we cannot reasonably assert that a true animal electricity is involved. These effects can and must be properly attributed to an artificial electricity excited at present by the new means indicated" (VO, 1:117: *Risposta alle domande dell'abate Tommasélli*, summer 1792).

But Volta's conclusion did not depend entirely on his experiments. The factual basis from which Volta induced the hypothesis that "it's the dissimilarity of metals that causes it" was largely the same as Galvani's. If the conclusion was different despite the nearly identical premises, this was due to other reasons. In reality, Volta's induction was guided not only by the factual premises, but by his acceptance and preference criteria—hence, by the underlying electrophysical assumption. It is chiefly this assumption, this gestalt, that drove him to a new theory of muscular contraction.

4.3 THE SPECIAL THEORY OF CONTACT ELECTRICITY

Volta did not wait very long to announce this theory. Even before publishing the second part of the *Memoria seconda*, he penned a brief paper for the Abate Carlo Amoretti's *Opuscoli scelti* (see Polvani 1942, 278ff). The text appeared anonymously in June 1792 under the title, *Transunto sull'elettricità animale ed alcune nuove proprietà del fluido elettrico*. It contains the first public statement of the theory of contact electricity: metals "should be regarded no longer as simple conductors, but as true motors of electricity, for with their mere contact they disrupt the equilibrium of the electrical fluid, remove it from its quiescent, inactive state, shift it, and carry it around. One metal, for example silver, will attract the fluid as if by suction; another, for example tin, will deposit it" (VE, 3:172).

In other words, the theory stated that metals have an electromotive power that resides in the contact between two dissimilar metals and a moist body. As we can see, the theory was typically phenomenological, in keeping with Volta's "second manner." Considering the specific function played by metals, we may call this the *special theory of contact electricity*. Using the distinction introduced by Volta (VO, 1:305; VOS, 454n) between first-class (or metallic) conductors and second-class (or moist) conductors, we can state the theory as follows:

V_s First-class conductors of different kinds have an electromotive power that is generated at their point of contact with the second-class conductors.

Volta's hypothesis for the specific mechanism of the imbalance was that one metal attracted the fluid and the other relayed it to the interposed moist body (fig. 4.3a). Volta gave an increasingly systematic exposition of his theory in his other writings later that year, particularly *Risposta alle domande dell'Abate Tommaselli*, the two letters to van Marum (summer), *Nuove osservazioni sull'elettricità animale* (November) and *Memoria terza sull'elettricità animale* (24 November). He also restated the theory in a few letters to Cavallo (1792–93).

The *Risposta* to Tommaselli remained somewhat conciliatory toward Galvani's theory. Volta began by recalling his main earlier experiment: the production of contractions by closing the arc on the nerves fitted with dissimilar armatures, leaving the muscles outside the circuit. The experiment showed that there was no discharge of the kind obtained with the Leyden jar: "And why seek one

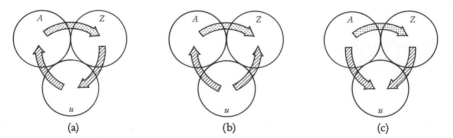

(a)   (b)   (c)

4.3 Three hypotheses on the electromotive power of metals according to Volta: (a) one metal gives electrical fluid to the moist body *u* (for *umido*), while the other metal receives electrical fluid from the body; (b) both metals receive fluid from the body to a different degree; (c) both metals communicate fluid to the body to a different degree. The hatched arrows stand for the main fluid flows; the dotted arrows for the currents that restore balance.

if there is no sign or need of it?" (VO, 1:114). But the conclusion was not yet completely negative: "Galvani's great discovery of a full-fledged animal electricity remains solid and stable; nonetheless, it must be limited to fewer phenomena, and nearly all his suppositions and explanations collapse."

Why Galvani's explanations should collapse is clarified in the following passage:

> When, for example, in order to obtain the contractions and movements discussed here, we must touch the muscle with one metal, and the nerve with another metal, or apply actually dissimilar armatures, whereas we cannot obtain the effect with identical armatures—then there is surely no reason to assume that a natural, organic electricity is at work here, and that a true imbalance exists in these parts, setting the electrical fluid in motion. Rather, one ought to say that if the fluid is shifted simply by applying two dissimilar metals, it is they that wrest the fluid from its state of idle imbalance, remove it from the places where they are applied, and carry it from one part to the other. . . .
>
> Metals are thus not only perfect conductors, but *motors* of electricity. . . . This is a new virtue of metals, which no one has yet suspected, and which I have been led to discover by my experiments. However, I believe it is specific not to metals alone, but in fact to all conductors; and I believe we must establish by a general law that the mere contact or juncture of conductors possessing different surfaces—and, more important, different qualities—is sufficient to disturb the balance of the electrical fluid in some manner and to move it without any need for rubbing. Such rubbing, as well as percussion and even mere pressure, prove all the more efficient for no other reason than for providing a better surface fit, by bringing a greater number of points into closer contact. (VO, 1:117; see fig. 4.4)

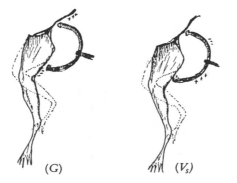

4.4 Galvani's theory (G) and Volta's special theory of contact electricity (V_s) compared. According to G: (1) nerve and muscle are naturally unbalanced; (2) the arc contact rebalances the fluid, producing contractions; (3) the arc has a passive function as conductor. According to V_s: (1) nerve and muscle are equally endowed with electrical fluid; (2) the arc contact displaces the fluid, whose current generates the contractions; (3) the arc has an active function as electromotor.

Volta's theory is clear; yet, compared with the one expounded shortly earlier in the *Transunto*, it contains a major innovation. The theory of contact appears here in a generalized form. The new statement, which we may call Volta's *general theory of contact electricity*, can be expressed as follows:

V_g All conductors of different kinds have an electromotive force generated at their point of contact.

Volta reiterated this theory in his second letter to van Marum (11 October 1792). Here Volta recalled an experiment performed on idioelectrics that consisted in "electrifying perfectly clean, dry glass and resin plates to a perceptible degree, setting them as delicately as possible on a mercury bath or on small cushions covered with equally dry and clean metal leaves, and detaching them just as gently" (VO, 1:136). Volta had discovered this property of idioelectrics "some years ago" while studying the Symmerian phenomena. At the time, he had used it to promote his theory of attraction, mentioning contact as a borderline case of rubbing and thus as a cause of imbalance (§2.1). Now, several years later, Volta went back to the experiment to use it as an indirect, analogical support for the explanation of the new experiments: as idioelectrics are electrified by contact, so conductors, when brought in contact, can effectively cause an imbalance.

Compared with the special theory, the general theory offered the undeniable advantage of greater parsimony. In particular, in compliance with Volta's preference criterion, it reduced the field of biological experiment to that of physical experiment; thus, it subsumed Galvani's electrobiological explanation under an electrophysical one by making muscular contractions a special case of electrical-fluid imbalance in conductors.

The fascination with parsimony was, however, insufficient, and in fact Volta did not feel too secure on this ground. His other acceptance criterion demanded direct proofs. But despite his repeated mentions of them (VO, 1:117, 136, 146) he possessed no proofs of such kind—least of all any independent proofs demonstrating a fluid imbalance even outside the frog. He therefore held to the special theory—apparently more secure from the experimental standpoint—and abandoned the general theory, at least for the moment. To explain the fluid-imbalance mechanism, he added two hypotheses to the preceding one (VO, 1:134, 212): (1) either both metals attracted the fluid from the moist body with unequal force, or (2) both repelled it with unequal force (fig. 4.3b, c). Gradually, he gave preference to the second one (VO, 1:379ff).

But even the special theory was still beset by "a few phenomena" that contradicted it, namely, "when the nerve is stripped and insulated in the manner of Sig. Galvani, and when we touch both the nerve and the muscle from which it emerges, using two tips of identical metal, or when both the muscle and nerve are armed with the same metal and in the very same manner, the convulsions are excited no less intensely" (VO, 1:117).

Volta's initial reaction was to reconcile the two theories—his own and Galvani's—by confining the latter to these few special cases. In his letter to Tommaselli, he wrote:

Yes, then we can indeed safely assert that the cause of such phenomena is a full-fledged animal electricity. And, in truth, since there is no reason why the electrical fluid should originate in the completely similar armatures, where can its motion come from, if not, at the outset, from the organic parts themselves to which the armatures are applied—the reason being that the fluid is unbalanced between these parts, that is, between the nerve and muscle, or between the inside and outside of the muscle in which the nerve penetrates and spreads? (VO, 1:117–18)

Very soon, however, even the conciliatory attitude vanished. In the already-mentioned *Nuove osservazioni* of November 1792, after reporting further experiments in the excitation of visual and gustatory sensations by means of dissimilar armatures, Volta dispelled any remaining perplexities.

The more I advance in my experiments, the more these doubts increase. Indeed, I am now convinced that no specific action of the organs, no vital force ever stimulates or moves the electrical fluid, or incites it to travel between different parts of the animal. Rather, the fluid is led and forced into doing so by an impulse it receives in the places to which the metals are joined—an impulse that drives it away and pursues it at one end, and pulls it at the other. I repeat, I am now convinced of this, especially when observing that nothing is ever, or nearly ever, obtained without the contact of a metal, or rather of two metals that are of different kinds or that are dissimilar in some other way, such as in hardness, smoothness, shine, etc. This leads me to conjecture that even when we obtain some convulsion and motion through the contact of two metals that seem to be totally identical (a very rare event, which occurs, if at all, only in the first moments after preparation, when the nerves' sensitivity is at its peak), the effect is due to some imperceptible difference between the metals.

If that is how things are, what is left of the animal Electricity claimed by Galvani, and seemingly demonstrated by his very fine experiments? Nothing other than the prodigious excitability of the nerves in charge of sensations and motion—especially voluntary motion—under the stimulus of the electrical fluid set flowing by external causes. This implies the nerves' merely passive disposition toward an electricity that is always extraneous, in other words, artificial. The nerves react to this electricity as simple *Electrometers*, so to speak; indeed, they are Electrometers of a new breed, incomparably more sensitive than any other Electrometer. (VO, 1:146–47; fig. 4.4)

In short, virtually nothing remained of animal electricity. Volta repeated this message openly a few days later, on 24 November 1792, in the letter to Giovanni Aldini that forms the *Memoria terza sull'elettricità animale*: "If that is how things are," in other words, if "the nerve alone is stimulated by the fluid traveling even on just a very brief segment of it, and if that is enough to excite the nerve into action and make it produce by itself (although in what manner, we confess we do not know) the contraction of the subjected muscle"—then "Galvani's theory and explanations, which You are striving to support, are largely disqualified, and the entire edifice is in danger of collapsing" (VO, 1:151–52; VOS, 424).

However, Volta was certainly too hasty in proclaiming the demise of Galvani's theory—a statement made, as Carradori put it (1817, 16), "with the thunder of truth." Volta was gripped by the urge to draw a conclusion from his own experiments and perhaps too by his pride as the great physicist who had found an explanation of an apparently biological phenomenon. In so doing, he stretched his own methodological criteria somewhat, and at any rate violated a series of wholly reasonable methodological criteria.

The first criterion he failed to observe was this: do not introduce untestable ad hoc hypotheses to save a theory disproved by experiment. The transgression was blatant. The special theory of contact electricity postulated that a flow of electrical fluid (hence the occurrence of contractions and sensations) was due to contact between metals "of different kinds." But this was disproved by experiment, since there were observable cases in which single-metal armatures, too, produced contractions and sensations. To salvage his theory, Volta had therefore introduced in the quoted passage of *Nuove osservationi* the hypothesis that such armatures could be "dissimilar in some other way, such as in hardness, smoothness, shine, etc." This was an auxiliary hypothesis, but also an ad hoc one, because *in principle* unfalsifiable. Indeed, the presence of the "etc." in the list of possible differences among metals[7] and the assertion, a few lines below, that the difference might also be "imperceptible," placed the hypothesis beyond all possible testing. It is obvious that if the two ends of a single-metal armature can always be said to differ on some nonobservable count, then there can be no operative definition of the concepts of identity and difference of contacts and armatures.

Another criterion also patently violated was the ban on circular explanations. By introducing the ad hoc hypothesis of imperceptible differences in the metals or in the armature's application points, Volta turned the theory of contact electricity into a question-begging exercise. On the one hand he stated that contractions occurred only in the presence of dissimilar armatures, but, when contractions did occur, he eventually claimed the armatures were dissimilar. Thus the explanation was circular, for lack of an *independent* criterion to define the dissimilarity of the armatures' metals or contact points.[8]

There was one last methodological criterion that Volta's theory challenged. Against rival explanations, Volta had advocated his acceptance criterion in terms of *direct* testing. But here the criterion turned against him. For a direct test of the contact-electricity theory required, among other things, that Volta provide evidence of electrical signs independent of the frog contractions. Despite his several references to such evidence (see p. 112) it is fairly doubtful he possessed it. In its absence, his theory remained little more than a plausible hypothesis.

Let us try, at this point, to make an initial assessment. After Volta's first objections, Giovanni Aldini had run for shelter behind an observation intended to weaken the falsification rule brandished by Volta against the animal-electricity theory. "If the good repute and integrity of scientific opinions," Aldini wrote, "were to be called in question whenever the slightest doubt was raised, we would certainly have few or no theories to regulate and guide the human conscience."[9] Now if we look at the state of the question by the end of 1792, when positions had already been clearly defined in both camps, we can see that the animal-electricity theory had a few experiments contradicting it (contractions with the arc on nerves alone, and contractions on whole animals only with dissimilar armatures) and a few experiments supporting it (contractions with single-metal arcs on prepared frogs). And while the contrary experiments, examined in the light of Aldini's remark, did not suffice to refute the theory, it is equally true that the favorable experiments were not sufficient to confirm it.

But the theory of contact electricity was in the same situation. It too had positive evidence (notably the contractions with the bimetallic arc on the nerve alone) and negative evidence (contractions with the single-metal arc). And if Volta could hope to overcome the latter by introducing an auxiliary hypothesis—a very doubtful prospect at that—he could certainly not turn the former into decisive arguments. Galvani had no independent proofs of animal electricity, because he had not proved that contractions occurred by joining animal parts *only*, without metal bodies. Similarly, Volta did not possess independent proofs of contact electricity because he had not proved that electrical signs could be produced by joining metals or conductors *only*, without animal bodies. In this situation, Volta's objection to Galvani—"the most plausible and attractive explanations, and also those that seem consistent with the initial, general appearances, are rarely confirmed by a more rigorous and more prolonged scrutiny of the phenomena" (VO, 1:57; VOS, 407)—could be just as properly turned against Volta's own theory of contact electricity.

Thus, at the start of 1793 and for the rest of the year, when the controversy between the two theories was building up, the situation lay at a standstill. The two camps stood face to face, each aware of its own strengths but also of its weaknesses. At first, reciprocal acknowledgments abounded. Gentlemanly gestures, in the manner of the time, were nevertheless accompanied by outspoken objections, as befits scientific practice. In reality, each camp, while apparently making concessions to the other, reasserted its own position. Volta, for example, continued to speak of Galvani's "fine" or "admirable" discovery (VO, 1:121, 174) while in fact rejecting it outright; Galvani wrote to Carminati that "a reasoning so correct, based on physical laws, and deduced from experiments by so accurate an experimenter" as Volta, "could not but deserve my genuine approval and a prompt change of my opinion" (GO, 143; GOS, 326–27), while at the same time restating his own theory without changing a comma; and Aldini, although conceding that Volta's opinion is "absolutely confirmed by recent experiments" (1792a, 232), continued to champion intrinsic animal electricity.

There was more to this theatrical role-playing than the scientists' pride, the stubbornness of strong personalities, or the protagonists' subjective advocacy of—and commitment to—their own explanations. Rather, in their initial confrontation, both theories were objectively no more than plausible hypotheses, each endowed with empirical support and credible theoretical anchors, but each beset with its own difficulties. In truth, Galvani's old dilemma—animal electricity or metallic electricity—although enhanced by new data, still awaited the decisive proof that would bring its solution. Alas, fate saw fit to distribute such proof equitably between both camps. As a result, each got its own crucial experiment.

CHAPTER 5

The Crucial Experiments

5.1 THE GALVANISTS' REACTION: ALDINI AND VALLI

The theory of animal electricity and the theory of contact electricity inevitably divided the scientific community. Although the closing years of the century were pregnant with momentous political processes and events, the vogue for electricity persisted and the connections between laboratories, learned societies, and *salons* remained intact. As du Bois-Reymond aptly put it:

> The storm aroused by the publication of the *Commentarius* among physicists, physiologists, and physicians can be compared only to the storm that at the same time [1791] arose on the political horizon of Europe. It can be said wherever there were frogs, and wherever two dissimilar metals could be fastened together, people could convince themselves with their own eyes of the marvellous revival of severed limbs. Physiologists thought they grasped in their hands their old dream of a vital force. Physicians—to whom Galvani had already supplied some tentative, summary advance explanations of mental illness, paralysis, tetanus and epilepsy—believed that no cure was henceforth impossible, and that no person was in danger of being buried if he had been previously galvanized. (1848, 1:50–51; see also Hoff 1936, 159)

Once Volta had published his *Memorie*, the opposing camps soon formed.[1] Simplifying the picture somewhat—for many authors adopted original, articulate positions—the following were among Galvani's supporters, or at any rate among Volta's critics: in Bologna, his nephew and research assistant Giovanni Aldini; in Pavia, Lazzaro Spallanzani, presumed author of the *Transunto* (partial translation) of the *Commentarius* and occasionally even a severe critic of Volta;[2] Eusebio Valli, a Pisan doctor who had made long stays in London and, independently from Galvani, contributed a decisive experiment to the controversy;[3] Antonio Maria Vassalli-Eandi, professor of physics in Turin, who had already developed a theory of animal electricity (VO, 1:21) and later questioned the identification of the electrical and galvanic fluids;[4] and the great Alexander von Humboldt in Berlin, who produced a theory of his own and steadfastly resisted Volta (VO, 1:539n).[5] The lineup of Volta's supporters—or, at least, of Galvani's critics—included: Giovacchino Carradori (see 1793), a Pisan doctor; Don Bassiano Carminati, professor of medicine in Pavia; Richard Fowler, in Edinburgh;[6] Christoph Pfaff of Stuttgart, who, at the same time as Volta (1793), had ranked metals according to their electromotive force;[7] Johann Christian Reil, doctor at Halle, who by late 1792 had formulated the hypothesis of electricity by contact between metals; and Friedrich Albrecht Carl Gren, professor of chemistry at Halle, who subscribed to the same hypothesis and developed it in the wake of his contacts with Volta (VO, 1:531–35). Felice Fontana, for his part, remained lastingly skeptical about animal electricity, while Tiberio Cavallo, in London—where he had gone on business and had settled to devote himself to science instead—never shed his initial perplexities on contact electricity (VE, 3:197; VO, 1:539n).[8]

Three successive events were to deepen the rift between the two camps: the crucial experiment in favor of Galvani's theory; the publication of Volta's reaction to it; and the crucial experiment in support of Volta's theory. Interestingly, though, despite the many reconsiderations and hesitations, there were no outspoken switches. Only those who until then had not committed themselves too visibly or had not played a leading role converted in 1794 to galvanism and, after 1798 or 1800, to Volta's ideas.

This widespread immobility may be largely due to the inertia typical of scientific communities and to their customary resistance to innovation. But psychological and sociological factors were not the only determinants. As we shall see, the conflicting positions were essentially conditioned by intrinsic reasons of merit and theoretical attitude. For the crucial experiments adduced by one side against the other always contained a residual element that not only withstood refutations but actually turned them into proofs. This element consisted of the conflicting interpretative theories of the phenomena, one electrobiological, the other electrophysical. It is not that the experiments were worthless or incapable of being expressed in each other's language. On the contrary, there was communication, compatibility, sometimes with decisive effects. Rather, under the pressure of the contending interpretative theories, the *same* experiments admitted of different explanatory hypotheses and theories. As a result, the crucial quality of such experiments was conditional on the prior acceptance of a particular interpretation, making the victories of either camp ephemeral, never definitive. However, each victory invariably forced the defeated camp to revise its position. This was particularly true for Volta after the Galvanists produced their crucial proof.

The Galvanists replied to the special theory of contact electricity within a seven-month period in 1794 with four critically important texts: Galvani's treatise, *Dell'uso e dell'attività dell'arco conduttore nelle contrazioni dei muscoli*, published in April;[9] Giovanni Aldini's *De animali electricitate dissertationes duae*, issued in the summer; Eusebio Valli's *Lettere sull'elettricità animale*, especially *Lettera XI* of 15 October 1794; and Galvani's *Supplemento al Trattato dell'arco conduttore*, which appeared in late October.

The attack on the special theory of contact electricity was based on two fundamental proofs. With a few well-conducted experiments, the Galvanists successfully showed that: (1) Contractions were also obtained with *homogeneous* armatures. This proof, provided by Valli, Aldini, and Galvani, was fairly convincing; however, its value as refutation was blunted from the outset by Volta's ad hoc hypothesis on the imperceptibility of the differences between the armature's contact points. (2) Contractions were also obtained *without* armatures, whether single-metal or bimetallic; this proof was furnished first by Valli and later, independently, by Galvani. It was unquestionably a crucial experiment because it refuted the theory of contact electricity between unlike metals and confirmed the theory of intrinsic electrical imbalance.

Let us examine these proofs in turn. The first was discussed in particular detail by Aldini, who tried to show that contractions could be produced even with a perfectly homogeneous arc. His experiment reproduced what he had added in an earlier-mentioned note to part 3 of the second edition of the *Commentarius*. Here is the description:

§IV. . . . The physicist Alessandro Volta, who excels in performing the highly ingenious experiments with which he is so familiar, raised very grave doubts in long letters that he kindly sent us first and published afterward in *Effemeridi ticinesi*. In truth, all his doubts were founded on this circumstance: heterogeneous metals are always required to produce contractions; some of the metals attract electricity while others give it back; and muscular motions are excited only when the dual electricity of metals is restored to balance.

§V. Who would fail to be captivated at once by so ingenious an idea, simply and elegantly presented? However, the experiments, together with those we performed ourselves, prevented us from adopting it wholeheartedly. From Galvani's *Commentarius* and our own repeated experiments it is clear that the frog's muscles and nerves, when immersed in two water-filled vessels, are sure to contract when approached by a metal arc. Here is a muscular contraction obtained with a single arc and a single metal. Live frogs, suffocated in a vacuum or in condensed air and dissected in the usual manner, exhibit a contraction without any armature, merely by the application of a silver arc to the nerves and muscles; and this is always the case, whether the frogs are large and strong or small and weak. Such an effect requires a balance between the two sides of a dissimilar armature: and in fact it is an identical arc that is brought up to the nerves and muscles. If, after observing the arc in its various components, we suspect it to be heterogeneous, why not extend that doubt, and suspect the countless other arcs with which we could excite the same contraction of being heterogeneous too? And if, among the many arcs employed, we find a single arc to be ever so slightly heterogeneous, then the excess and deficiency of electricity are to be found not in the dissimilar metallic matter but in the various parts of the animal. While these ideas were being entertained, Galvani informed our Academy that he had obtained contractions in dissected fresh frogs without the aid of the pneumatic machine, without armatures, but simply by applying an arc. The same happened to Carradori, who, after some initial doubt and

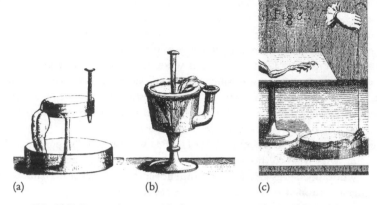

5.1 Aldini's experiments with the mercury arc. From Aldini 1794.

hesitation, concluded that the phenomenon was constant and proven. For our part, although our observations tally with Galvani and Carradori's, we find that the action of the pneumatic machine, even without the armatures, seems to excite electricity in frogs to a further degree.

§VI. However, to make our experiments an even more solid proof of animal electricity, and as it proved difficult to find a solid metal that a scrupulous chemist would regard as thoroughly homogeneous, we resorted to a fluid metal, namely mercury, and with several chemical processes we made it so pure that nothing better could be desired. To enable the mercury to serve reliably as armature and arc, we devised a few apparatuses that, although all intended for the same purpose, so differ in their use as to require separate descriptions. Two glass vessels are placed one inside the other [fig. 5.1a]; the upper vessel, filled with mercury, receives the spinal cord of a prepared frog and is fitted with a hole at the bottom; this can be opened freely and lets the mercury drop on a part of the frog's muscles below. When this happens, a contraction arises; but the arc consists of mercury, as does the armature; therefore the electricity is the same in both; consequently, there is no reason to expect an action from external electricity. As a result, when we observe a contraction, we must not argue that it derives from metallic electricity. But might we not imagine that the mercury, in its fall, extracts electricity from the glass walls that it strikes, just as the upper part of a barometer glows with a resplendent electrical light when the mercury is shaken even lightly? Whoever fears that the effect is due to the glass vessels need only use wooden ones and he will immediately realize that his apprehensions are scarcely justified.

§VII. The use of this machine presents certain drawbacks, which, without proper attention, may be blamed on a defect of the machine, whereas they are due to the person who maneuvers it inaccurately. For the spinal cord, because of its lightness, may sometimes remain on the mercury's surface when it should instead be immersed. To prevent this, we must use an arc made of glass or another nonconducting body to press the cord so it remains completely immersed in the mercury. If this step is neglected, we will observe certain irksome anomalies that will jeopardize the success of the experiment. To avoid them more easily, take a glass siphon [fig. 5.1b] with two arms, one larger, one narrower, the inner walls in the form of an inverted conical surface ending in a funnel hole that can be opened at will. Pour some mercury down the narrower arm of the siphon; it will flow into the larger arm but not into the inverted conical surface unless one opens the funnel. After setting up the device thus, fill the narrower arm with the spinal cord of a frog immersed in mercury; and, while its muscles are in the empty conical surface, open the hole. As soon as the mercury returns to equilibrium it will invade the muscles, and a contraction will be observed; this is due to the formation, from the spinal cord to the muscles, of a mercury arc highly suited to contractions.

§VIII. But to arrive at the proposed experiment in the easiest manner, we used a simpler procedure. Take a glass vessel filled with mercury [fig. 5.1c] and let the muscles of a prepared frog float on the surface of the metal. Hang the spinal cord by a silk thread at a certain distance from the mercury's surface, in such a manner that it can be freely brought closer to the surface by lowering the thread. When this is done, a contraction occurs that will be just as easily produced if we replace the mercury by a gold or silver plate—the latter metal, however, having less power to transfer animal electricity. These phenomena occur not only in a whole frog, but in a frog divided lengthwise halfway down the middle, which, if dipped in the mercury as described above, will contract sharply. (1794, iv–vi; for the contemporary English translation, see Aldini 1803, 136–40; Fulton and Cushing 1936)

These experiments were quite convincing. In a letter of 10 July 1794 to Galvani, Spallanzani said that, together with those contained in the *Trattato*, they "confirmed your discovery so wonderfully, and throw such light on the phenomena and on the laws governing it, that I would be unable to name another physical truth that could be so forcefully demonstrated" (1964, 5:55; see also 1976, 124–25). Humboldt was of the same opinion, but more realistic about the experiments' value as refutation:

> Aldini in Bologna has opened a surer path; he has publicized the experiments with mercury and superseded all his forerunners in the subject owing to the variety and refinement of his experiments and the ingenious method he has adopted. However, because his findings tended to overturn a generally accepted theory to which people were strongly attached, he met the same fate that awaits all scientists in similar cases: the facts he adduced were denied and he was accused of error. (1799, 45)

As a precautionary measure against Aldini's experiment and all others of the same kind, Volta had effectively advanced the hypothesis that seemingly identical metals in fact differ because of imperceptible qualities. But, as we have already remarked, Volta's move was typically ad hoc. In any event, Aldini also took up and won Volta's other challenge by producing contractions not only with homogeneous metal arcs, but with *non*metallic conducting arcs. For this purpose, Aldini substituted carbon for silver and obtained the same results: "Why then ascribe to the different power of metals, effects which can be produced by bodies which certainly have nothing of the metallic quality?" (Aldini 1794, xvi–xvii; English from Aldini 1803, 153).

The conclusion could then be set out in a few points:

1. It is not necessary to use unlike metals to generate contractions; a single one suffices; silver, and especially gold, are excellent for more resistant animals.

2. Any lingering suspicion of heterogeneity in solid metals can be easily eliminated by using a liquid metal, namely, mercury purified through chemical processes.

3. A contraction is produced when one armature and the arc are made of mercury, and when the arc, in its flow, does not strike the muscles; thus, there is no need to attribute the phenomenon to a stimulus effect, which many experiments demonstrate to be nonexistent.

4. When no effect is to be expected from the external electricity of the armature and arc naturally or artificially in balance, nevertheless an animal electricity is produced that controls the contractions.

5. Lastly, it seems that no metallic electricity can be suspected when the arcs and armatures are of charcoal. If no metallic elements are employed, there is no reason to argue that they are the source of animal electricity. (Aldini 1794, xvii; for the contemporary English translation, see Aldini 1803, 153–54)

But there was a second, even more convincing proof against the theory of the electromotive force of metals. It was Eusebio Valli's experiments against the theory that made the strongest impression on Volta. Valli first related them in *IXe Lettre sur l'Electricité animale*, dated 2 December 1792—just a few days after Volta, in the *Memoria terza*, had announced to Aldini that Galvani's theory was collapsing. After having personally confirmed it, Valli rehearsed the old galvanian opinion that "to excite shocks in the freshly killed frog, a single metal conductor is enough. It is unnecessary to arm either the nerves or the muscle" (1793b, 74). This experiment—it will be recalled—was initially accepted even by Volta, although he insisted that a certain class of contractions required a bimetallic arc. But, the following year, Valli produced a new experiment that disproved even this opinion of Volta's—indeed his entire theory. In *Experiments on Animal Electricity* (1793), Valli reported that "at two or three different times I had produced shocks by being myself the conductor" (1793a, 38), that is, by establishing communication between the frog's nerves and muscles via his body *without* resorting to metallic conductors.

Finally, in *Lettera XI*, Valli reviewed the entire question in order to supply "new ideas" on "the latest literary differences arising between Signori Volta and Aldini" (1794, iii). Here is a typical experiment of his:

After placing a frog on a surface (I am talking about prepared frogs: I shall not specify this again) I press my left thumb on the upper part of the animal's thighs, while with my right hand I fold back one of its legs on its spine, forming a sort of arc. Each time I touch it, the frog jerks, leaps and, I'm tempted to say, escapes me. The jerks and leaps become less frequent and less forceful as the animal expires, and their total duration does not exceed thirty minutes.

The experiment is not successful with sickly or small frogs, and sometimes fails even with frogs that appear to be full of vigor and highly excitable. (Pp. iv–v)

Experiments such as these, variously repeated, left no room for doubt. Indeed, Valli's conclusion was highly explicit and critical:

We observe motion in frogs that are no longer alive, by establishing communication between the muscles and their respective nerves. Therefore electricity in a state of imbalance exists in the muscles, in the nerves, or in both. We observe motion without the aid of metals. Therefore metals are not the motors of electricity. It is not they that cause imbalance. They possess no secret magic virtue. (P. xviii)[10]

In other words, in the face of such experiments, one could no longer continue stating that "it's the dissimilarity of metals that causes it." The findings were clear: the diversity of metals had nothing to do with it, because their alleged action was also performed by nonmetallic conductors. "From this moment on," Valli proudly concluded, "animal electricity ceases to be a problem. I am honored to have contributed to the triumph of a discovery that is the finest and most interesting of our age" (pp. xviii–xix).

These words, Valli's experiments, and those of Aldini and Galvani must have had a deep impact. Carradori, who militated for Volta and was equally shaken by these new findings, aptly described the ensuing climate: "These facts and arguments in favor of galvanism had a great success. So deep was their impact on the minds of Physicists that many who had declared or were about to declare themselves for Volta's system reverted to Galvani's. The Galvanists' ranks swelled to such a point that they once again encompassed nearly everyone. All scientists predicted imminent defeat for Volta, and total triumph for Galvani" (1817, 20).

Volta was not defeated but, on his own admission, suffered a setback. He reacted almost immediately but had to run for shelter by amending his theory. Before examining his next move, however, we should take a closer look at the objections voiced against his theory of the electromotive force of metals by the protagonist of the rival camp—Galvani himself.

5.2 GALVANI'S "THIRD EXPERIMENT": CONTRACTIONS WITHOUT METALS

The aims of the treatise *Dell'uso e dell'attività dell'arco conduttore* (hereafter *Trattato*) are clearly set out in the preface. Galvani defined four: (1) to demonstrate the "necessity and activity" of the arc—of any arc—in muscular contractions;

(2) to prove that the contractions were due to electricity; (3) to prove that such electricity was inherent in the animal; (4) to advance "some conjecture" as to the natural arc and therefore the contraction mechanism. Broadly speaking, Galvani sought to avoid a situation where scientists would "found theories on untrue principles, thereby delaying the advances in physiology that these new courses of experiments seem to promise, or following paths that would lead one away from the sought-for truth" (GO, 156; GOS, 366).

Unlike the *Commentarius*, the *Trattato* is constructed systematically. It is not an account of experiments in chronological order—as in the *Commentarius*—but a presentation of experiments arranged in a logical order to confirm a theoretical concept. In both texts, however, the procedure is typically inductive, relying on both generalization and abstraction. Galvani starts with the experiments, derives the empirical law that there are no cases of contraction "in which the arc's action is not manifestly observable," reaches the theoretical conclusion that "all muscular contractions . . . depend on a highly subtle fluid, an electrical one" (GO, 194; GOS, 385), and finally broadens the conclusion with the notion that this electrical fluid is intrinsic. Thus, moving to ever wider levels of generalization or deeper levels of abstraction, Galvani proceeded from muscular contractions to animal electricity.

The systematic, compact nature of the text calls for an analytical scrutiny. We shall limit ourselves here to the first three points; the fourth will be discussed at the appropriate moment (§6.4). Let us begin with the first point—that all contractions require an arc. For a start, Galvani observed, the arcs can be classified. The main classification, according to the quality of their component matter and the number of their component parts, comprises four categories:

Type A: simple, homogeneous, one-part arc
Type B: simple, homogeneous, multiple-part arc
Type C: compound, dissimilar, one-part arc
Type D: compound, dissimilar, multiple-part arc

Galvani went on to observe that the different types of arc indicated the "varying degrees of the animal's natural force" (GO, 159; GOS, 368). Here we must also note the argument's theoretical underpinning—a total reversal of Volta's assumptions. For Volta, muscular contractions measured the electromotive force of the arc; for Galvani, it was the nature of the arc that revealed the force of animal electricity. The experiments were the same, so much so that the degrees of the frog's force according to Galvani corresponded, at least in part, to the "degrees" of the frog's death according to Volta.[11] But the different theoretical selectors introduced them into different domains and imposed divergent explanatory hypotheses. Galvani did not doubt that the domain of his phenomena was the biological one, and that the explanation had to be electrobiological. Conversely, Volta did not question the principle that the domain and the explanatory hypotheses had to be physical and electrophysical.

However, although ultimately decisive, the selector had no impact on the first aim of the *Trattato*. Galvani demonstrated the need for the arc simply by observing that there were no cases of contraction that did not require some type of arc. Sometimes the arc effect was not evident, in which case one was dealing with a true "natural arc."

Nor did the electrobiological selector exert a bias on the second aim, which Galvani reached through an inductive reasoning sufficiently neutral—and, from a historical standpoint, sufficiently well established—to win acceptance even from those who, like Volta, had other theoretical commitments. The argument ran as follows. What could one "deduce from Galvani's new experiments" (GO, 194; GOS, 385)? *Two* things, immediately. First, that contractions depend on a "very subtle fluid"; second, that this fluid is electrical. The first conclusion seemed obvious, for the contractions were due to "an agent that operates instantly, that in producing the effect takes a certain determined circular path through certain determined bodies, both fluid and solid, even hard and very compact bodies. Manifestly, therefore, such an agent cannot be anything else than a very subtle fluid" (ibid.). The second conclusion was just as obvious. The electrical nature of the fluid "may apparently be deduced from the following fact: the same bodies that let the electrical fluid through—and are therefore called conductors, such as metals, waters, etc.—also let this fluid through; and the same bodies that block the electrical fluid—and are therefore called nonconductors, such as glass, resin, oil, etc.—similarly block this fluid" (ibid.).

Galvani speaks here of "deducing," but the term, in keeping with seventeenth-century Newtonian usage, corresponds more accurately to an induction, specifically, an analogical induction. Galvani's argument basically ran as follows: if the "essential characteristics" of the nervous fluid determined from the experiments are identical to those regarded by the physicists as sufficient to define an electrical fluid, then—by analogy—we are entitled to conclude that the muscular fluid *is* the electrical fluid itself, or at least that the hypothesis is "proper and reasonable." Thus Galvani reverted to his old strategy of using the findings of physics to support physiology. The argument was unchanged: if physics is correct—in other words, if it is accurate in inferring the existence of an electrical fluid from a class of typical observations—then the same physics must also accept the two following conclusions: (1) muscular contractions are due to a fluid; (2) this fluid is the electrical fluid itself.

But could one also conclude that the electrical fluid is *inherent* in the animal? This was the third and most important aim of the *Trattato*. And it was here that the electrobiological selector became decisive. In examining this point, the *Trattato*'s approach was even more systematic. Galvani did not reply directly to the objections against his theory. He proceeded by "retracing the matter from its principles" (GO, 196; GOS, 386). First, he presented the logically complete spectrum of possible hypotheses; then, he refuted them one by one. His demonstration, however, was not a proof by exclusion.

Galvani began by recalling that "the existence of an action and motion dependent on electricity requires an imbalance in the latter" (ibid.). The range of logical possibilities therefore comprised three hypotheses: "That this imbalance exists in the conducting bodies applied to the animal, in the animal itself, or between the animal and the bodies." These three hypotheses—let us call them H_1, H_2, and H_3—can each be subdivided. If the imbalance is between the conductors, we have four distinct hypotheses—H_{1a}, H_{1b}, H_{1c}, and H_{1d}—depending on which of the four arcs A, B, C, and D is applied to the animal. If the imbalance lies instead between the animal and the conductors, we have two separate hypotheses, H_{2a} and H_{2b}, depending on whether the positive element is the animal or the applied conductor. Lastly, if the imbalance is inherent in the animal, the specific hypotheses are again four—H_{3a}, H_{3b}, H_{3c}, and H_{3d}—depending on whether the imbalance is regarded as natural in the animal, induced in the animal by dissimilar metals or conductors, induced by the different states of the parts of the animal to which the arc is applied, or induced by the differences in application time. The first case is that of animal electricity; the second is that in which "the animal must be regarded as a generic moist body, as Signor Volta argues" (GO, 202; GOS, 388); the third is the "ingenious conjecture" of Carradori (1793, letter 5, p. 6); the fourth is Caldani's hypothesis. (See the table for a list of the complete range of hypotheses examined by Galvani.)

H_1 Imbalance between the conductors applied to the animal	H_2 Imbalance between the conductors and the animal	H_3 Imbalance inherent in the animal
H_{1a} Imbalance between homogeneous, one-part conductors	H_{2a} Imbalance between a negative animal and positive conductors	H_{3a} Natural animal imbalance
H_{1b} Imbalance between homogeneous, multi-part conductors	H_{2b} Imbalance between a positive animal and negative conductors	H_{3b} Animal imbalance induced by the dissimilarity of metals or other conductors
H_{1c} Imbalance between dissimilar one-part conductors		H_{3c} Animal imbalance induced by the difference between the animal's naked state and its armed state
H_{1d} Imbalance between dissimilar multi-part conductors		H_{3d} Animal imbalance induced by the diversity of arc application times

Galvani refuted all the H_1 hypotheses with an identical kind of argument. Let us consider H_{1a} and take, for example, a silver arc:

Either it is completely uniform in substance, uniform in shape, polish, and size, and in that case it cannot be reasonably assumed to possess imbalance; or, as Signor

Volta claims, it is not uniform, and even then such an imbalance seems impossible to suppose; on the contrary, whatever the said lack of uniformity, it is ultimately necessary for the electricity of one part to strike a balance with the electricity of the other part, and consequently throughout the body, since the parts that compose such a piece of metal are bonded together and conductive. (GO, 196–97)

The same argument applies to the other types of arc and to the conducting bodies; thus the hypotheses H_{1a}, H_{1b}, H_{1c}, and H_{1d} are to be regarded as all equally refuted. Drawing on one of the most firmly established laws of electrical science, Galvani stated: "It can generally be asserted that the electrical imbalance whose force lies at the origin of the above-mentioned contractions is not to be found in the conducting bodies that are applied to the animal" (GO, 198; GOS, 387).

Also refuted, for Galvani, were the objections to the H_2 hypotheses. Here his arguments are of various kinds, based partly on experiment and partly on intuitive considerations of plausibility. Let us take H_{2a}. There are six objections to it. If the contractions were due to an imbalance between the positively charged conducting bodies and the negative animal, then

1. To excite contractions, it would be sufficient to have "just one contact of the above-mentioned bodies with any single point of the animal: but a single contact between them in a single point, even in a nerve, is never sufficient to excite contractions; this requires two simultaneous contacts in two distinct points of the animal" (GO, 198).

2. After the animal is touched and "total, free communication of the electricity is obtained, everything should be in balance, and no further contraction should arise from new contacts; despite this, the contractions are repeated ten, twenty, thirty times, even if the animal and arc are insulated" (GO, 199).

3. The "imbalance would be removed by handling the metal arc as many times as the experiment is performed and repeated; yet despite these explicit and, I would say, almost infinite contacts, the contractions are obtained just as well as when the arc is used with insulation" (ibid.).

4. It should also be impossible to elicit contractions "when the arc is immersed either in a nonconducting fluid like air or in a conductor like water" (ibid.).

5. If it were objected that contractions are obtained when the metals can acquire new electricity from the atmosphere, one could not explain how they are also produced when the animal and the arc are immersed in oil (ibid.).

6. Finally, since contractions are obtained even in large animals with "two pieces of metal foil, one tin, the other brass, both of them square, each measuring only one Paris line [1/12-inch thickness]," there would be an "enormous disproportion between these effects and the supposed reason" (GO, 200).

Therefore it was "beyond doubt that an overabundant electricity in these conducting bodies is not the one that, by flowing into the animal to achieve balance, excites the said contractions" (GO, 201; GOS, 387). Reasons (1) and (6) also disqualified H_{2b}. Galvani thus concluded that "it seems clear and proven that the imbalance in question does not lie between the conducting bodies and the animal either. We must therefore seek it in the animal itself" (GO, 201; GOS, 388).

This brings us to the H_3 hypotheses. Galvani immediately tackled H_{3b}, which was Volta's theory. And if he believed he could refute it without citing any new experiment, it is because the many experiments performed by himself and Aldini with single-metal arcs must have seemed totally convincing to him. Galvani merely recalled that since

> a single, simple metal, well smoothed all over its surface, such as pure, simple mercury applied to the bare parts of the animal . . . and . . . in addition two small pieces of the tin foil applied to the crural nerve of a recently killed calf, at two opposite points, and connected by their loose ends . . . produce the contractions continuously, two legitimate consequences follow: first, that such contractions are not accidental; second, that neither the diversity and heterogeneity of metals, nor the capacity of one to displace electricity and of the other to attract it, are the true and efficient cause of the phenomenon of contractions, as these are fully obtained without them. (GO, 202–3; GOS, 389)

As for H_{3c}, Carradori's hypothesis, Galvani found its refutation in the experiment where contractions "are obtained by means of the same very simple homogeneous arc, the frog's spinal cord being placed in a water-filled vessel and its legs in another totally identical one. In this case the armatures are identical and consist of a fluid that attracts and conducts electricity to the same degree as the animal fluid, or certainly to a not very different degree" (GOS, 204; GOS, 389).[12]

The same fate awaited H_{3d}. Caldani's hypothesis, said Galvani,

> although on the face of it very plausible, loses much force and plausibility when one considers, first, that the contractions are produced by dividing the dissimilar arc composed only of tin and brass leaves, whose tips were both applied to the animal shortly before; second, that to obtain the contractions one need only change the contacts between the leaves that form the arc, the change taking place simultaneously with respect to both leaves. (GO, 205; GOS, 390)

Thus, after all the possible or actually proposed alternative hypotheses had been refuted, the only choice left was to admit the single survivor, H_{3a}: "As a result," wrote Galvani, "one must confess that it [the imbalance] is naturally indwelling in the animal. But this could not be so if the animal were a simple generic moist body. We must therefore recognize the existence in the animal of a special machine capable of producing and conserving such an imbalance naturally" (ibid.). This machine, however, "remains totally concealed from the most penetrating scrutiny." Consequently, "we can perceive only its properties, and from them, somehow, conjecture its nature" (GOS, 390). In what manner? Once again, Galvani's instrument was analogy. The animal machine exhibited three relevant properties: (1) that of "containing two contrary electricities"; (2) that of "keeping the two electricities basically and constantly divided and insulated"; (3) that of "clinging so fast to its electricity as to prevent it from escaping." And "since these properties are also the same characteristics and basic features of the Franklin square or Leyden jar," then it is reasonable to compare the animal machine to such a jar.[13]

In this conclusion, the weight of the electrobiological selector is evident. For Galvani, an *electricity in the animal* is an *animal electricity* precisely because the animal is an animal, a living organism, not a "generic moist body." In the last resort, *ceteris paribus*—that is, when both contestants had admitted the existence of an imbalance between parts of the animal—the decision to regard the imbalance as a sign of animal electricity revealed by the arc, or of common electricity generated by the arc, depended on how one regarded the body in which the electricity manifested itself. Therefore, the decision depended on the interpretative theories involved.

We shall have occasion to return to this sensitive point of the entire controversy. Here, it remains to be added that Galvani did not intend to prove his theory only indirectly, by exclusion or by analogy. The substantial novelty of the *Trattato* is its report of a *crucial experiment*—known as Galvani's "third experiment"—in favor of animal electricity and against the theory of contact electricity. The experiment—a very famous one, the same as Valli's—was the following (fig. 5.2):

> To dispel any doubts, therefore, and also any suspicion of deliberate contrivance, we shall propose another experiment, considerably easier and simpler, and even more reliable and convincing, which I successfully performed a number of times last summer in the presence of learned physicists. The experiment is as follows. Prepare the frog in the same manner. Cut the crural nerves near their entrance into the spinal cord. Then, without immersing them in any solution or modifying them in any way, let them hang immediately from the prepared ossicle, as in the first experiment. Next, bring into contact, as before, the lateral parts of the thigh, either by lifting them with a nonconducting body and letting them drop on the thigh, or by pushing them with the nonconductor into a slight mutual contact, and, if possible, in a single point of the muscle. The moment the contact takes place, both legs will be seen to contract, and—to one's surprise—the leg that was previously dangling will jerk up slightly. Such contractions are obtained almost always four and even more times in any freshly prepared and reasonably healthy frog, and more often if it is of above-average size—the experiment succeeding more easily in this case owing to the greater nerve length. It should further be noted that the phenomenon also occurs, albeit sometimes more sluggishly, when only one of the two nerves is put in contact with its corresponding thigh—in which case the contractions, obviously, are obtained in that part alone. Not only is the experiment easier and simpler than the first, but it is also—and this is of greater interest—decisive, in my opinion. (GO, 211–12; GOS, 395–96)

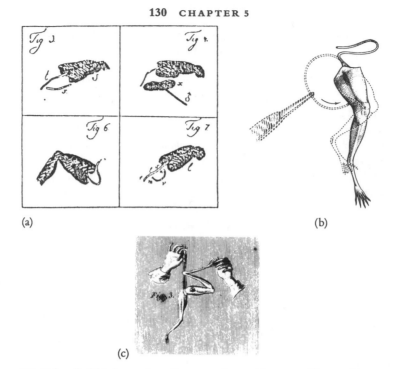

(a)   (b)

(c)

5.2 Galvani's "third experiment": contractions without metallic arcs. From
(a) Humboldt 1799; (b) Sirol 1939; (c) Aldini 1804.

Even if the experiment was not "decisive," as Galvani believed and others
later claimed (see La Métherie 1802, 27; Medici 1845, 19; Fulton 1926, 36;
Geddes and Hoff 1971, 41; Rowbottom and Susskind 1984, 45), its probative
strength was certainly impressive. It demonstrated precisely what Volta's theory
denied. But there was more. In Galvani's view, the experiment had the added
advantage of reinforcing the analogy between the Leyden jar and the innervated
muscle. Here the muscle exhibited a new, hitherto unobserved property, corresponding to an already known property of the jar:

> All of these facts and the differences indicated here constitute—in my opinion—
> new, quite clear and convincing proofs of the following proposition: the electricity
> that causes the contractions induced by Signor Galvani's devices and our own is
> utterly different from common, extrinsic electricity; it is an electricity entirely spe-
> cific to the animal, in which, of necessity, it must remain collected and distributed
> in a machine highly analogous to the Leyden jar. This seems demonstrated by the
> last experiment even more vividly than by the others—since by placing the nerve
> on the thigh one produces the same effect that would occur in the jar if its hook
> were bent back to touch the outer surface. Also, in the same experiment, the con-
> tractions do not occur when the nerve is put in contact with the muscles of another

animal; similarly, no effect would occur if the hook of one jar was brought in contact with the outer surface of another jar. (GO, 214–15)

Now confident of his own experiments and of their interpretation, Galvani went one step further. The entire tenth chapter of the *Trattato*, devoted to "reflections on some experiments and some objections," is a detailed critique of the supposed proofs of the metallic-contact theory. In particular, Galvani refuted Volta's experiments on the excitations of nonmotor nerves, that is, on sensations. To begin with, Galvani sternly remarked, the experiments were so ambiguous as to make them unfit for consideration by any serious scientist: "To deduce . . . the nature, action and direction of a fluid's motion from the taste it excites will always be a highly uncertain and highly equivocal process; hence, an inference as to the nature of the electricity that acts on the tongue . . . will not have the force and truth of conjecture—and even less so of ratiocination—that we desire from philosophers" (GO, 235–36). Moreover, "even if one granted" Volta "that such sensations were constant . . . would it legitimately follow that the electricity was of the common sort, activated by metals?" Hardly, because the experiments show that such sensations are obtained only when the arc is placed on a person or on organs. If they were due to the common electricity communicated from the outside, what would be the purpose of the arc (GO, 236–37)?

In *Nuove osservazioni sull'elettricità animale*, Volta had written: "If that is how things are, what is left of the animal Electricity claimed by Galvani, and seemingly demonstrated by his very fine experiments?" (VO, 1:147). And now Galvani used almost the very same words to rebuke his opponent: "But if that is how things are—if such electricity is indeed wholly specific to the animal, and not common and extrinsic—what will become of the opinion of Signor Volta, who, with the experiments he adduces, has claimed to rule out animal electricity entirely, and to restrict Galvani's discoveries to the sole revelation, in the animal, of the most sensitive electrometer known?" (GO, 237). This opinion of "Signor Volta" was bound to fall[14]—and so it did, but not in the way Galvani expected.

5.3 VOLTA'S MOVE: THE GENERAL THEORY OF CONTACT ELECTRICITY

The objection against crucial experiments is well known. Roughly speaking, it argues that an experiment contrary to an observational conclusion logically deduced from a theory does not falsify the theory itself. This is because, in practice, the conclusion does not follow from the theory alone, but from a complex consisting of the theory, auxiliary hypotheses, and initial conditions. As a result, the negative experiment strikes at the *entire* complex, not just at a single part of it. On grounds of principle, the criticism is impeccable. In point of fact, however, although it is true that a theory under test is always inserted in a complex of hypotheses, one can often identify or reach a consensus as to which of the hypotheses in the complex is responsible for the falsification.

Such was the case with Volta. As we shall see, he came to recognize that his *special* theory of contact electricity—the theory that "it's the dissimilarity of metals that causes it"—had been genuinely disproved by Galvani and Valli's experiments. However, far from regarding these contrary experiments as devastating, he cleverly managed to transform them into evidence for his *general* theory of contact electricity—which he promptly revived for the occasion. We must now look at this turnaround and this theory.

Reduced to its essentials, the criticism leveled by Aldini, Galvani, and Valli against Volta was as follows. Let us assume that—in addition to the known laws of electricity—Volta was right in claiming that the muscular contractions of the frog are due to an electrical current generated when unlike metals are joined with a moist conductor (the frog's muscle). Let us further suppose that the frog is placed in certain specific experimental conditions: for example, the nerve and muscle are touched with a perfectly homogeneous metal; the armatures are applied in the same manner; the nerve alone is joined with the muscle. In these cases, contractions should *not* occur. Now, since the contractions *do* occur in the experimental circumstances indicated, and on the assumption that the laws of electricity are valid, it follows that Volta's theory is falsified by experiment. In symbols, this complex argument is summed up by the formula:

$$(V_s \cdot (L_1 \cdot \ldots \cdot L_n)) \supset ((C_1 \cdot \ldots \cdot C_n) \supset {\sim}E) \cdot (E \cdot (C_1 \cdot \ldots \cdot C_n) \cdot (L_1 \cdot \ldots \cdot L_n)) \supset {\sim}V_s.$$

V_s is Volta's special theory of contact electricity. $L_1, \ldots, L_n$ are the set of laws of electricity and the hypotheses taken for granted during the test. $C_1, \ldots, C_n$ are the set of experimental conditions that, singly or jointly, should produce the contractions (for example, C_1 = joining of nerves and muscles with a perfectly homogeneous metal; C_2 = application of the armatures in the same manner; C_3 = simple contact between nerves and muscles). E are the contractions.

This reasoning was quite convincing—and indeed it persuaded many. As Carradori wrote in a passage already quoted: "All scientists predicted imminent defeat for Volta, and total triumph for Galvani" (1817, 20). Now, to counter the Galvanists' argument, Volta had three alternatives available:

   1. To admit E and divert the falsification to $C_1, \ldots, C_n$, that is, to raise objections to the conditions in which his adversaries had performed the experiments
   2. To admit E and $C_1, \ldots, C_n$ and divert the falsification to $L_1, \ldots, L_n$, that is, to attribute the failure to some of the hypotheses related to V_s
   3. To admit E and $C_1, \ldots, C_n$, preserve $L_1, \ldots, L_n$, and regard V_s as falsified

In this last case, Volta could either

3a. Regard V_s as refuted—that is, disproved by experiment—and abandon it; or

3b. Regard as refuted a particular version of V_s and therefore preserve the theory in a different version.

Volta never seriously envisaged the second course—nor, for that matter, did anyone else in the controversy. Initially, he adopted the first, but after Valli's *Lettera XI* appeared, he admitted his error and resolutely took up the third. Of the two options that (3) still allowed him, he chose the second. His move consisted in saving his theory—or at least its core—from falsification by reformulating it. Although the move was largely ad hoc, Volta's new theory was propelled by new facts that enabled it to resist and, ultimately, to establish itself.

As we know, the special theory of contact postulated that in order to obtain a fluid imbalance, hence contractions, the two metals joined to the moist conductor had to be dissimilar. For in the opposite case—if both ends of the conducting arc in contact with the frog's nerve and muscle were entirely identical—then their electromotive actions would also be identical. And since these would be perfectly balanced, they would not produce the fluid circulation that alone could elicit the contractions. But the dissimilarity requirement was precisely Volta's first line of defense against the Galvanists' offensive: in other words, Volta raised objections to the three experimental conditions C_1–C_3 in which his opponents had obtained contractions. Volta set out his triple rebuttal in the "Lettera seconda" to Vassalli-Eandi in *Nuova memoria sull'elettricità animale* (1794).

To Aldini's experiment, Volta objected that it did not satisfy the condition C_1, that is, the perfect homogeneity of the metal used as conducting arc; even rectified mercury, according to Volta, exhibited differences among its parts, "and great ones indeed between its internal parts and its superficial parts, which promptly lose their luster in the air and suffer from incipient calcination, especially when shaken" (VO, 1:276n; VOS, 442n). Hence "mercury . . . , which Aldini selects as the most reliable, is on the contrary the most suspect and unreliable."

Similarly, Volta objected that condition C_2 required to make Aldini's experiment conclusive—namely, that "armatures of the same metal . . . must also be applied in the same manner"—was not unequivocally fulfilled. Volta's reasoning here is quite interesting, because it shows to what extent one can divert the experimental falsifications of a hypothesis: "Now if this has not been observed in the experiments produced against me, if such a perfect identity has not obtained there too, I *can always say* that one of the two pieces of metal, owing to a difference in its application, has prevailed on the other, despite its being of the same kind and otherwise identical" (VO, 1:276n; VOS, 443n; italics mine).

Lastly, according to Volta, not even condition C_3 indicated in Galvani's crucial experiment—the simple, direct juncture of nerves and muscles—was reliably fulfilled. It is true, Volta admitted, that "here . . . there is no metal, nor any other conductor, nor armatures, nor arc. Or rather, the arc from nerves to muscles is formed by the muscles and nerves themselves. Therefore the electrical charge and discharge exists, and occurs in the animal parts alone. Consequently the electricity is specific to organs, not accidental and external." And yet, Volta added, these consequences were nowhere near as rigorous as they seemed; they "would be acceptable if the experiment was dependable and consistent, and if there were not a lingering doubt—indeed a very grave suspicion—that a mechanical irritation was the cause of the effect in such proofs, rendering these wholly inconclusive" (VO, 1:280n; VOS, 449–50n). In fact, the experiment was not always successful;[15] as for the irritation, Volta went on:

> Do not tell me that, when one part is gently tipped toward the other, so as to join together rather than strike each other, no perceptible mechanical irritation seems to arise. For I reply that it is not so easy to prevent the nerves or stump from falling on the thighs and clinging to them with some force. This is due to the superficial humidity that draws those parts together through mutual attraction, with the result that they join with an accelerated motion. Such a phenomenon is not hard to observe; and if we observe it closely we shall also see that the convulsions do not occur unless a blow or impact takes place; and on a very few occasions they happen even in its presence, the other favorable circumstances such as extreme sensitivity, etc., being sufficient, as I have explained. (VO, 1:281; VOS, 450–51)

Volta regarded his replies as excellent, and this argument in particular as "one of the most victorious." But it must be admitted that some of his objections are specious and contrived, and that in developing them he went overboard, "gripping the razors by the blades—indeed, climbing up the mirrors"—as Polvani recognized (1942, 307–8). Was this a manifestation of the gestaltic or magical effect sometimes induced by theories?[16] Or was it more simply an instance of the bias exerted by the polemical spirit, which, when not properly controlled, prevails over the precision of experiments and the evidence of observational data? (We would be loath to suspect a hint of stubbornness and bad faith in an experimenter so accurate that he later even recognized his mistake.) The fact remains that Volta went much further. Not only did he reassert his theory to the full, but, discarding all precaution, launched into apodictic assertions that exposed him to the serious risk of experimental disproof. For example, he rashly stated:

> Apart from metals, ores (some of which are very rich or very poor in metal, and I have even found some pyrites equal to ordinary metals) and charcoal—which, for the property in question, must be put in the same category as metals—no other conductor is capable, when applied as armature, of producing the electrical taste on the tongue, or the coruscation of the eye, or the burning sensation, or any movement in the muscles of the liveliest frog, even the best prepared. (VO, 1:274; VOS, 441–44)

At other times he threw down the gauntlet:

> And why, if the electrical fluid is unbalanced in the animal organs. . . , do not the same motions occur; why does the frog remain perfectly still when, instead of using metals, we form the arc with other good conducting bodies such as a rope, wood, a piece of cardboard or other bodies—moist, wet and dripping with water—or when we use two fingers, or even when we dip both hands completely, one in each glass? (VO, 1:274–75; VOS, 444–46)

Volta felt so sure of himself that he even affixed a methodological stamp to his conclusions: "One should stick to the facts," he wrote at the beginning of the "Lettera seconda" in his *Nuova memoria* to Vassalli-Eandi, "and to their immediate consequences without venturing much further and indulging in conjectures and hypotheses that are not entirely based on them" (VO, 1:271; VOS, 437). Yet it was the facts that betrayed him. He had quibbled with Galvani's crucial experiment. But when it was ably repeated by Valli—who even alluded ironically in his conclusion to the "secret, magical virtue" ascribed by Volta to metals (§5.1 above)—Volta could no longer resort to expedients. He honestly admitted that the experiment contradicted his theory. He did not concede, however, that it was a total refutation. Of the two options then open to him—to reject his theory or amend it—he chose the second. In a key passage of the third letter to Vassalli-Eandi (7 October 1795), Volta wrote:

> I don't quite know how you reacted, first, to the experiments of that kind performed by the Bolognese Professor [Galvani]; then to my far more extensive and varied ones, from which I drew very different consequences utterly unfavorable to the supposed animal electricity; and finally to Valli's new experiments and other similar ones, with which some have claimed to re-establish it on unshakable grounds. I know that the latter experiments have impressed many, as I said before. These observers have seen convulsions somehow occur in freshly prepared, highly sensitive frogs, even without the intervention of a metallic or carbon conductor—a phenomenon that I had ruled impossible, because I had been previously unable to produce it (and in fact it is difficult to obtain). As a result, these observers searched no further before conceding victory to the advocates of animal electricity in the strict sense—that is, of the alleged charge or imbalance of electrical fluid between the nerves and their corresponding muscles, or between the inside and outside of those muscles. Yet such experiments, as I shall undertake to demonstrate, are no proof of animal electricity. They only show that I went too far when I stated that by applying moist (i.e., second-class) conductors alone—that is, without using any metal or first-class conductor—one could never excite convulsions in frogs however they are prepared and even if they are highly sensitive. It is on this point that I must retract, in other words, correct my excessively general statements. But this does not impair my fundamental proposition, which I have argued and continue to argue, that the electrical fluid does not receive the impulse from the animal organs

in which—as the Galvanists contend—the fluid finds itself charged or unbalanced. Rather, the fluid receives it from a force generated by the contact between unlike conductors that enter the circuit. In other words, even in such experiments, in which metals are not used, it is an artificial electricity excited by an extrinsic cause, an external agent; and in no way by a principle or force inherent in the animal organs, nerves and muscles. (VO, 1:291; VOS, 455–57)

Strictly speaking, therefore, Valli's experiments constituted both a falsification and a confirmation for Volta. They were a falsification, because they proved that muscular contractions could be obtained even without first-class conductors. This was a cogent, unobjectionable circumstance that totally invalidated a prediction Volta had confidently derived from his theory and brandished against his opponents. But the experiments were also a confirmation, because they proved that contractions always required contact between dissimilar *conductors*, even if they were only of the second class.

The falsification applied to the *special* theory of contact electricity. By contrast—and far more importantly—the confirmation reinforced the theory's core, its "fundamental proposition": that the electrical-fluid imbalance was not inherent in the animal body but caused by an *external* force. The end result was that Volta managed to recover his *general* theory of contact electricity, according to which the electromotive force was generated by contact between first- and second-class conductors, *as well as* between second-class conductors alone, provided the conductors were dissimilar in either case. To revamp Volta's old and effective phrase, it was no longer "the dissimilarity of metals that causes it" but "the dissimilarity of conductors [that] is required" (VO, 1:293; VOS, 459).

Volta found evidence for this requirement, and hence for the new formulation of his theory, in the very experiments cited against him. Both Valli and Galvani's experiments showed that the contractions occurred, or were facilitated, when (1) the frog's body was daubed with blood or moistened with saliva (Valli) or wetted with salt water (Galvani);[17] (2) the parts of the frog brought into contact were dissimilar. Actually, Volta exaggerated the first circumstance, because in the experiment of the *Trattato* that Galvani considered "decisive," Galvani explicitly said he had cut the crural nerves and brought them up to the muscle "*without immersing them in any solution*" (GO, 211; GOS, 395; italics mine). However, it was true that salt water revived the effect when it was about to cease.[18] At any event, in Volta's opinion, the circumstance was sufficient to ascribe the contractions not to a natural imbalance between nerve and muscle, but to an electromotive force generated by the contact between two conductors (nerve and muscle) in a circuit of three conductors (nerve-muscle-liquid), all second-class and dissimilar (fig. 5.3). As Volta wrote to Vassalli-Eandi, in the same "Lettera terza,"

both conditions are therefore required: (1) the interposition of a dissimilar liquid; (2) the dissimilarity, so to speak, of the animal parts that face each other; in other

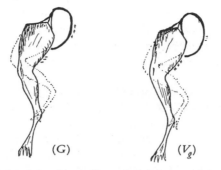

(G)   (V_g)

5.3 Galvani's animal-electricity theory (G) and Volta's general theory of contact electricity (V_g) compared. According to G: (1) nerve and muscle are naturally unbalanced; (2) their contact restores the fluid balance, producing the contractions. According to V_g: (1) the contact between nerve, muscle, and interposed liquid (humor) generates an electromotive force; (2) the electrical fluid set in motion flows in a loop, producing the contractions.

words, these must not be too similar—such as muscle and muscle—or, even less, identical in structure and consistency; on the contrary, they must be considerably different. The difference I find most propitious to conduction is precisely the one between tendon and muscle or between tendon and nerve (the one between muscle and nerve is not so effective). As a result, the difference that works best between these animal conductors and the third conductor (which must come between them when the contact is made and the circuit is closed) is obtained when this third, mediating body is a viscous or saline liquid, or better a distempered soap, or better still a barely liquified alkali, as I have already indicated. (VO, 1:296; VOS, 463)

He then concluded: "If you consider and reconsider the question from every angle, this is the only way to explain such experiments and countless others, which come down to the same principle, as I will show" (VO, 1:297; VOS, 465). Whether this was truly the "only way" is doubtful; indeed, we can be certain that it was not so at all. Yet Volta's explanation was unquestionably ingenious and plausible. To begin with, it was not an improvisation. Volta had already formulated the general theory of contact in his letters to Tommaselli in August 1792 and to van Marum in October 1792 (see §4.3 above). While we may be skeptical about the motives for his subsequent abandonment of it,[19] we must admit that Volta now objectively appeared to be rehearsing previously exposed ideas.

Regarding the foundation, however, the situation is different. Certainly, the general theory of contact electricity met a major methodological requirement: the parsimony of explanations or unification of principles:

If these few, ambiguous experiments can be . . . sufficiently explained by the sole principle of the action of dissimilar conductors—a principle demonstrated by so many other clear, eloquent and incomparably more numerous experimental proofs—why resort to another alleged and unproven principle, namely, an active electricity specific to animal organs? Why introduce two utterly different principles for wholly similar phenomena of the same kind? (VO, 1:293; VOS, 459)

Moreover, Volta went on, Galvani's theory was beset by "new difficulties and inexplicable anomalies" (VO, 1:307). One of these was that the contractions were weak or even absent when homogeneous parts of the frog were brought into contact, whereas they occurred normally when the contact was formed between dissimilar parts, or even between like parts but with "a multiplication of imperfect conductors." Now while this phenomenon was understandable in the context of the theory of contact electricity, in Galvani's theory it produced "an inexplicable paradox with respect to our notions of electrical charges, and of conductors simply considered as such, that is, as bodies permeable to the electrical fluid, and to nothing else" (VO, 1:296–97; VOS, 464).

But even Volta's theory contained an awkwardness—regarding the very dissimilarity of contacts—that made it objectively weak and suspect. Gliozzi criticized Volta for "(arbitrarily) inverting the law of contact. Instead of saying 'in the contact of dissimilar bodies, an electrical imbalance manifests itself,' he stated 'if we encounter an electrical imbalance, this means there is a contact between dissimilar bodies'" (1937, 2:122). This is like saying that Volta's theory was ad hoc and his proposed explanation was circular. Indeed, the same objections raised against the special theory of contact electricity can be leveled at the general theory. For lack of other checks than the experiments it purported to explain, the theory was truly ad hoc. Also, as it lacked an independent criterion for the dissimilarity of conductors, it was *circular*. From the *rhetorical* standpoint, Volta might well feel invulnerable, since, at least as regarded certain parts of the frog (nerves and muscles) he could derive their dissimilarity from Galvani's own statements. But from the *logical* standpoint, he was on weak ground, because the dissimilarity had to be proved and he could not do so without introducing effects other than electrical ones (the contractions).

Another circumstance helped to make the general theory of contact electricity unfalsifiable. Even if Volta had introduced independent criteria of heterogeneity, such as physiological rather than electrical ones, the fact remained that the causes he listed for heterogeneity always ended with an "etc." and were not rigorously enumerated and defined.[20] As a result, he could always rebut a disproof by brandishing some hidden property or, as Galvani put it, "subtlety." The term occurs in the *Trattato* when Galvani, in anticipating a few such rebuttals, remarked on the subject of salt water that "as the advocates of dissimilar armatures . . . could not deny the identity of the fluid, they might deny the identity of the effect in the nerves and muscles; and supposing the same sea-salt altered the nerves in one manner and the muscles in another, they might perhaps manage to defend their opinion by regarding these different alterations as different armatures" (GO, 213). Thus, Volta's "etc." functioned as the theory's self-immunizing clause. And Volta himself—as Polvani, his most sympathetic supporter, recognized—"could not fail to realize that the weak spot of his entire construction was the need to use the frog, the tongue, the eye or other organic systems for his experiments. Until that condition had been eliminated, one could always argue that the origin, the prime cause of the observed phenomena lay right there, in the organic system" (Polvani 1942, 321).

In short, the two contestants could hope to gain the upper hand by dint of crucial experiments. Galvani had now produced his, but Volta not yet. In such circumstances, he could hope at best to cast doubts on the rival theory, but not to prove his own. This does not mean, however, that Volta's program had lost its drive or—in the language of Imre Lakatos—had become "regressive." The general theory of contact electricity did not confine itself to fitting the facts, to merely adapting itself to new phenomena by means of suitable alterations.[21] It also predicted *novel* facts. Specifically, it predicted that electrical signs would be displayed by combining dissimilar first-class conductors and even mere second-class conductors. Thus Volta opened a new pathway for research and prepared his crucial experiment. In so doing, he was able not just to ward off his opponents, but also to counterattack. He would have won, at least from his point of view, if he had succeeded in finding the right combination of dissimilar conductors, as well as the way of physically measuring the electrical current they should have produced—that is, a measurement independent of the frog's contractions or of organic phenomena.[22]

5.4 VOLTA'S CRUCIAL EXPERIMENTS

In his quest for a crucial physical proof, Volta systematically tested the circuits formed by three or more dissimilar conductors. He began his examination in the two letters to Giovanni Mocchetti (June and August 1795) and in the fourth letter to Vassalli-Eandi (20 December 1795); he reported his further findings in the first two letters to Gren (August 1796). At the same time, Volta tried to develop a ranking of second-class conductors as he had done several years earlier for first-class conductors.[23]

Let us therefore consider the various possible circuits, starting with those formed by two conductors. A simple reasoning will show that they are not sufficient to cause an imbalance: "When the two bodies *aabb* [fig. 5.4a] are joined together, the action of the electrical fluid produced by such junctures must be identical on both sides—that is, the fluid will tend to flow from *a* to *b* or from *b* to *a*, from either side in equal measure. Consequently, whatever the action, such contrary forces or tendencies counterbalance each other, so that no current can be generated either from right to left or from left to right" (VO, 1:322–23). This was confirmed by experiment:

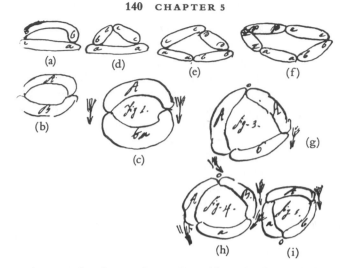

5.4. Contact between first-class conductors (capital letters) and second-class conductors (lower-case letters). Cases *h*, *i*, and *d* are the three basic experiments of the Galvanists, respectively with a bimetallic arc, with a single-metal arc, and without a metal arc. From VO, 1:323, 326, 379, 323, 323, 323, 380, 381, 381.

If *A* [fig. 5.4b, c] is also a metal, provided it is identical at both ends not only in kind, but in temper, polish, etc., and *B* the body of a frog, provided the parts on which the said bodies of arc *A* rest are similar, the frog, however excitable, will exhibit no commotion. The reason is that, while the union between a metal and a moist conductor strongly impels the electrical fluid, the impulse is exerted from both sides in opposite directions and with equal strength, preventing the formation of a fluid current. (VO, 1:326)

The contractions were obtained, however, by introducing a third body (fig. 5.4d) or a fourth body or more (fig. 5.4e, f),

belonging to either class, provided it is of a different species from the one of the homologous class; since if it was also of the same species and entirely identical [fig. 5.4g] . . . the two would be equivalent to a single continuous one. . . . That is not so if the piece introduced differs from both the others—namely, from one as to class, and from the other (which is of a homologous class) as to species, as in [fig. 5.4h, i]. For although here too the actions oppose each other, as shown by the arrows, they are not equal and balanced, since there is a difference in the degrees of force resulting from the juncture of *A* and *a* on the one hand, and from the juncture of *B* also with *a* on the other hand in [fig. 5.4h]; similarly, there is a difference of force in [fig. 5.4i] between *A*'s juncture with *a* at one end and with *b* at the other end. (VO, 1:379–81)

In other words, contractions were obtained when, owing to the presence of a fluid imbalance, an electrical current was generated: this took place when "one action prevails on the other, precisely in proportion to the degrees of prevalence" (VO, 1:381). In particular, the imbalance occurred in the Galvanists' three canonical experiments. Fig. 5.4h, in which *a* is the frog, illustrates the ordinary experiment with the bimetallic conducting arc—the experiment or "manner . . . that has been common practice from the beginning in 1792 to today in the wake of Galvani's discoveries" (VO, 1:401; VOS, 480). Fig. 5.4i (where *a* and *b* are the frog's nerve and muscle) illustrates Galvani and Aldini's experiment with the single-metal arc. Finally, fig. 5.4d depicts the experiment "produced so triumphantly by the Galvanists and especially by Doctor Valli" (VO, 1:411) in which the circuit is formed by muscle, nerve, and blood or liquid. To sum up, there are

three ways to incite the electrical fluid and make it flow, producing all those convulsions and sensations; all three manners or combinations come down to introducing at least *three different conductors*: first, *two metals, or first-class conductors of different kinds*, which are in direct contact on one side and communicate on the other by means of one or more *moist*, that is, *second-class conductors*; second, *a single metal* in contact *with two different moist conductors* communicating with each other; third, *three moist*—that is, *second-class—conductors*, all *different*. (VO, 1:417; VOS, 428; see the synoptic table in fig. 5.5)

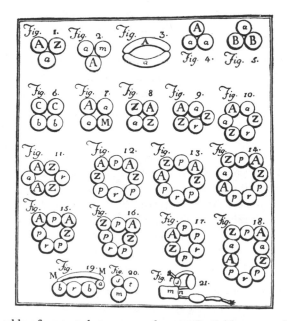

5.5 Synoptic table of contacts between conductors. Capital letters stand for first–class conductors, lower-case letters for second–class (moist) conductors: muscle (*m*), nerve (*n*), and blood (*s*) (for *sangue*). One can recognize the three basic experiments of the Galvanists in figs. 1, 2, 21. From VO, 1:398.

Among the possible combinations, one case was left to be examined: that of three first-class conductors. Does such a circuit produce an electrical current, Volta asked?

> Analogy seems to suggest so, and the principle would gain by becoming extremely general, to wit, that the electrical fluid is made to circulate whenever and however one forms an unbroken circle of three different conductors. But this cannot be demonstrated in the absence of an electroscopic body, so to speak, capable of indicating it—that is, the prepared frog; and also in the absence of other nerves and muscles excitable even by a weak current: these bodies belong to the second class of Conductors, and are lacking in the case envisaged here, in which the circuit is composed only of first-class bodies. (VO, 1:377)

The principle was not only incapable of being demonstrated. It also appeared to be refuted because, Volta wrote to Mocchetti, "other experiments seem to prove that the contacts between first-class bodies, however these may differ in kind, are incapable or virtually incapable of inciting the electrical fluid and causing it to form a current. Above all, we observe that if we take two similar or dissimilar metals, one touching a second-class body on the right, the other to the left, and if we insert a third metal, and even many other different ones, between the two, there is no notable change in the effects" (VO, 1:377). But Volta was not yet aware that this experiment *confirmed* the principle he was denying.

In other words, so far—and it was now August 1795—there were still no genuine proofs, that is, independent of the frog's contractions, to support the general theory of contact electricity. Indeed, whatever proofs there were seemed to contradict it. But Volta did not give up. Exactly one year later, in August 1796, he reexamined the combination of two metals. Obviously Volta's most cherished concept—the electrophysical assumption or expectation—was in jeopardy. If electricity came from metals, why did these not reveal it directly?

This time the expected results did materialize. Between August 1796 and March 1797 Volta discovered "some new facts" and made "even clearer and more eloquent experiments" that revealed in a physical manner, without organic intermediary, the electromotive force of a pair of unlike metals. Volta performed some of these experiments with Nicholson's doubler; others, "all the clearer and more decisive for being simpler" (VO, 1:435; VOS, 498), with his own condenser. Here is a very famous one, carried out with the condenser, as reported by Volta in 1801 and still described today in every physics textbook (fig. 5.6; see Figuier 1868, 694–95):

> §V. After setting up a good condenser of this kind, I perform the following fundamental experiment. I bring into contact a piece of pure silver or silver alloy, such as a coin, and a piece or small strip of zinc, or I join them together with a screw, nail or metal fastening of some kind, or in any other way, provided the two metals are in contact: I join them together in the manner indicated, or in any other manner I prefer. Taking the piece of zinc z in my fingers, I touch the other piece

5.6 The electromotive power of metals: Volta's experiment with the condenser. From Figuier 1868.

of silver *a* briefly to the upper plate of the condenser, while the lower plate communicates, as it should, with the ground. I then withdraw the pair of strips *az*, and raise the upper plate, which has received the electricity of the silver strip *a*. (The electricity has gathered and accumulated in the plate in proportion to the plate's capacity and collecting power when coupled with its companion. This is due to the contrary electricity the companion acquired when grounded, as postulated by the familiar theory of the condenser.) When I do this, the plate displays an electricity by default (El. −) of 2 to 3 degrees, and sometimes even up to 4 on my thin-straw electrometer, as the plate shows when brought near the electrometer and placed in contact with the latter's head. (VO, 2:52–53; VOS, 543–44)

The conclusions to be drawn from such an experiment were beyond doubt, and Volta clearly spelled them out to Gren: after rehearsing the three previous hypotheses (see §4.3 and fig. 4.3), he introduced his fourth and definitive one regarding the mechanism that produced the fluid imbalance when dissimilar metal conductors came in contact with a moist body (fig. 5.7):

> The contact between, for example, silver and tin gives rise to a force, an exertion, that causes the first to *give* electrical fluid, the second to *receive* it: the silver tends to release it, and releases some into the tin, etc. If the circuit also contains moist conductors, this force or tendency produces a current, a continuous flow of the fluid, which travels in the above-mentioned direction from the silver to the tin, and

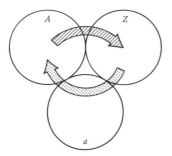

5.7 Volta's definitive hypothesis on the electromotive power of metals: one metal emits electrical fluid, the other receives it; the moist conductor restores the balance.

5.8 Contact electricity. "The three disks of the duplicator are made of brass; I take two rods, one of Silver ["*A*" in the drawing] the other of Tin ["*S*"]. I apply the former to the moving disk, the latter to one of the fixed disks, while both rods rest on the table, or better on a wet cardboard, or other moist conductor ["*a*"]—in other words, they communicate by means of one or more second-class conductors. After leaving the apparatus in this state for a few hours, I remove the two rods and set the machine in motion. After 20, 30, 40 revolutions—or more, if the air is not dry, or if the insulators are in poor condition—when I touch one of my thin-straw Electrometers to the moving disk, I read very distinct signs of *plus* Electricity (+*E*), up to 4, 6, 10 degrees and even more. When I touch it to one of the fixed disks instead, I also get signs, but of contrary, that is, *minus* Electricity (−*E*)" (VO, 1:468).

from the tin via the moist conductor(s) back to the silver and then back to the tin, etc. If the circuit is not complete, if the metals are insulated, the result is an accumulation of the electrical fluid in the tin at the expense of the silver—in other words, a positive or *plus electricity* in the former, and a negative or *minus* one in the latter. Admittedly, the electricity is weak, and below the degree required to move ordinary electrometers. Despite this, I was finally able to make it more perceptible than I ever hoped. (VO, 1:419–20; VOS, 484)

Yet this type of proof remained inadequate. To establish the general theory of contact electricity, a second and third type were needed. Volta had to obtain electrical signs through contact not only with (1) first-class conductors alone, but also with (2) first-class and second-class conductors together, and (3) with second-class conductors alone. This time, however, the second-class conductor had to be a generic moist body—not Galvani's frog.

Even with the second type of proof, Volta was successful. In a memoir of 1796–97, addressed to van Marum but never published, he reported numerous experiments that all basically consisted in putting two unlike metals, joined by a moist body, in contact with the doubler's two fixed disks. The result was that the moving disk exhibited distinct signs of a contrary electricity (fig. 5.8).

The third type of proof, however, turned out to be more complicated. Did Volta succeed in obtaining electrical signs simply by combining moist generic bodies? None of the 1796–97 experiments described attests to the fact, although Volta claimed success on two occasions (VO, 1:552, 2:32n). In all likelihood, though, such experiments never proved entirely conclusive. Once Volta had produced electrical signs from contacts between first-class conductors only, and between first- and second-class conductors, he must have extrapolated the case of purely second-class conductors from the results already arrived at by using the frog as electroscope (VO, 1:377–78). Therefore, he probably never secured what would have been the most decisive proof. However, as the next chapter will show, this failure did not and could not influence the outcome of the controversy.

At any rate, the results Volta had obtained were, objectively, considerable—indeed, for him, they were much more: they were *crucial* experiments. On the strength of them, he had no more misgivings—assuming he had ever harbored any—in asserting that an electrical current is generated from an imbalance produced by the contact between dissimilar conductors, and that such imbalance is "very great" in the contact between first-class conductors (VO, 1:555); "middling" in the contact between first- and second-class conductors (ibid.); "of scant efficiency," "weaker," or "fairly small" in the contact between second-class conductors (VO, 1:378, 552, 555).

The experiments truly seemed to speak for themselves. What more could one have desired? On 1 August 1796, when the general theory of contact electricity was still more a project than a reality, Volta concluded his first letter to Gren with these words:

> Herein lies the whole secret, the whole magic of *Galvanism*. It is simply an artificial electricity, which acts there under the impulse of contacts between different conductors. It is these that, strictly speaking, function as the true prime *motors*: nor is such a virtue exclusive to metals, or first-class conductors, as one might have believed; it is common to all conductors, to varying degrees, depending on their nature and fitness—even, to some extent, to moist or second-class conductors. Follow these principles and you will clearly explain all the experiments carried out so far. You will not have to resort to another imaginary principle of a specific, active animal electricity in the organs. On the contrary, you will invent new experiments and predict their success, as I have done and continue to do every day. Give up these principles or lose sight of them, and you will no longer find in such a vast field of experimentation anything but uncertainties, contradictions, endless anomalies—and everything there will become an inexplicable enigma. (VO, 1:413)

By the spring of 1797, at the end of his new cycle of proofs, Volta's position was—if anything—stronger. Like Galvani, he now possessed his worthy crucial experiments. It was a curious circumstance, which should have led both men to reconsideration and compromise. Interestingly, reciprocal offers of compromise were in fact volunteered, but it is perhaps even more interesting to understand why they were rejected.

CHAPTER 6.

The Pile

6.1 MISSED COMPROMISES

In 1797, a year before his death, Galvani took up his pen for the last time to draft the five *Memorie sull'elettricità animale* to Lazzaro Spallanzani. Their purpose was to refute Volta's theory "with new considerations and new attempts." At a certain point, however, in the second memoir, he examined what obviously appears to be a compromise hypothesis. This consisted in admitting *two* types of electricity—"animal electricity" and "common electricity"—and consequently two causes of contractions in the frog, a natural one and the one induced by "artifices." "Such a supposition would have made Signor Volta's opinion very compatible with mine, and one would not have destroyed the other" (GO, 324; GOS, 440).

When Galvani wrote this, he was still unaware of Volta's new experiments on contact electricity between dissimilar conductors, because the letters to Gren that reported them were not published until 1797, the same year as the *Memorie* to Spallanzani. "There is every reason to believe," wrote Gherardi in a note on the second memoir, "that if he had not been lost to science right after these major steps by his famous opponent toward his greatest discovery, [Galvani] would have espoused the assumption he put forward in this section—an assumption regarded as not implausible" (GO, 324–25n; GOS, 440n).

It is hard to say if things would really have turned out that way. Indeed, there are good reasons to doubt it. At all events, that is not what happened. "I have found such difficulties in the assumption indicated," Galvani added, "besides those set out in the anonymous Work [his own *Trattato dell'arco conduttore*], that I have been incapable of resolving them and, therefore, of subscribing to its apparent plausibility" (GO, 324; GOS, 440). Thus the divergence deepened into an "all or nothing" split. Listing his disagreements with Volta, Galvani wrote in the first *Memoria*:

> He wants this to be the same electricity that is common to all bodies; I want it to be special and specific to the animal. He locates the cause of the imbalance in the contrivances used, particularly in the dissimilarity of metals; I, in the animal machine. He determines such a cause to be accidental and extrinsic; I, to be natural and internal. He, in short, attributes everything to metals, nothing to the animal; I, everything to the latter, nothing to the former, as far as the imbalance alone is concerned. (GO, 303–4; GOS, 429)

Before examining the difficulties Galvani put in the way of a compromise, it is worth glancing at his detailed reply to Volta's move. From the theoretical standpoint, the *Memorie* to Spallanzani contain no innovations; from the experimental standpoint, they are noteworthy for another famous experiment—the "fourth experiment," as we may call it—and the observations on torpedoes. Among these, one in particular reinforced Galvani's electrobiological conviction: when the torpedoes had their hearts removed, they continued to give signs of electricity; the signs, however, ceased immediately if the fish were decapitated. Galvani saw this as a reliable indication of the animal electricity's organic nature and of the brain's role as a storage and sorting center for the electrical fluid.

But the fourth experiment was a confirmation—indeed a major one. As seen in chapter 5, Volta explained Valli and Galvani's experiments by attributing the contractions obtained without metals to the contact of dissimilar second-class conductors. Galvani initially objected that this explanation was equivalent to his own ("If one adds that the dissimilarity also requires the circuit described, who can fail to see that we're both saying the same thing, albeit with different words?" GO, 322; GOS, 438). He then performed an extremely clever experiment to remove all lingering doubts. The experiment and the conclusions Galvani drew from it deserve to be quoted in full.

> I am well aware of the objection one could perhaps raise against these facts—namely, that however much it may be granted that the arc and armatures are homogeneous, or lack the necessary dissimilarity, nevertheless, when we establish contact in all these experiments between the nerve of the prepared animal and a muscular substance, we are always bringing into contact two totally unlike bodies. As a result, it would be argued, these experiments do not rule out the force of dissimilarity. In truth, such an objection loses most of its thrust after the experiments indicated above, and the comments on them made just earlier. Still, to destroy this shadow of doubt even more completely, I conducted the following experiment [fig. 6.1]. I

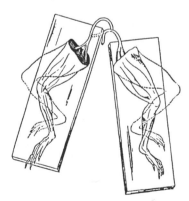

6.1 Galvani's crucial "fourth experiment." From Sirol 1939.

prepared the animal in the usual manner. I cut each sciatic nerve near its exit from the spinal cord. I then split the legs apart from each other, leaving each with its corresponding nerve only. I folded the nerve of one leg into a small arc, raised the nerve of the other leg with the usual small glass cylinder and dropped it on the nerve arc. In so doing, I made sure the falling nerve touched the folded nerve in two points, one of these being the nerve orifice. I saw a movement in the leg whose nerve I dropped on the other; at other times, I saw both legs move. The experiment succeeds when the legs are totally insulated and do not communicate with each other except through the contact of their nerves. . . . Now what dissimilarity could be called in to help explain the contractions that occur here, since the contact is formed between the nerves alone? Perhaps some will speak of the stimulus felt by the nerves in falling on each other. But then why do we fail to obtain contractions if we shake one of these same nerves on a much harder and rougher arc made of non-conducting material—for example, of sulfur or glass? After all, the stimulus generated by the impact should be considerably greater in such a case. It seems to me ascertainable, therefore, that there exists another series of contractions obtained without stimulus, without metal and without the slightest hint of dissimilarity; they are thus produced by a circuit of electricity inherent in the animal and naturally imbalanced in it. (GO, 322–23; GOS, 439)

There is no doubt that the experiment was, on its own grounds, of the highest significance.[1] Indeed—within the limits we shall spell out later—it was decisive, because it showed the presence of an electrical current at least in one case categorically excluded by Volta: the contact between two moist conductors that were homogeneous according to his criterion (see fig. 5.4a and fig. 5.5, nos. 20–21). Spallanzani was probably referring to this experiment when he wrote that in the five papers dedicated to him, Galvani "among other things replies to the objections of my very illustrious colleague Don Alessandro Volta, and, in my view, reduces them to dust" (1964, 180).

On the other hand, although Galvani did not yet know about them, Volta's experiments on the electromotive force of dissimilar conductors were just as decisive on their own grounds and within the limits to be examined later. A compromise on a distinction of fields and principles would therefore have been a reasonable solution. Why then, after having envisaged it, did Galvani promptly reject it? Why didn't Volta accept it either? And why was another compromise offered by Volta a year later equally spurned?

First, let us look at Galvani's reasons, the most important of which can be reduced to three:

1. "It is hard to understand how the electrical imbalance induced by the diversity of metals and by some of their slight differences could be so great and strong as to produce muscular contractions despite tremendous obstacles" (GO, 325; GOS, 440). In other words, Galvani saw a disproportion between the cause (the contact between two small dissimilar strips) and the effect (the contractions obtained even through a long, obstacle-ridden circuit, such as when the arc "is formed in the ground").

2. The "constant experiment" (GO, 329; GOS, 442) of contractions with homogeneous, metallic, or simply conducting arcs.

3. Very early on, Volta had shown (see sec. 4.2) that "contractions are obtained [even] with the application of the arc and dissimilar armatures to the same nerve"— a "very forceful argument for excluding the two contrary electricities, that is, the existence of a natural imbalance in the muscle." Galvani even confessed that this experiment had nearly led him to "abandon my opinion for Volta's" (GOS, 339; GOS, 447). And he might have done so had he not come across other experiments that "totally undermined Volta's arguments" by showing (a) that the same contractions are obtained also by applying homogeneous armatures and a single arc to the nerve; (b) that they are obtained also by applying the armatures and arc to a detached nerve and putting it in contact with the nerves and muscles of a prepared frog (GO, 340), thereby excluding direct irritation; (c) that no contractions are obtained in the same experiment if the detached nerve—complete with armatures and arc—is placed perpendicularly on the frog's nerve, thereby ruling out the action of electrical atmospheres (GO, 341). The experiments showed that even in this case the electricity circulates from muscles to muscles via the nerve and arc. (For instance, in fig. 6.2, nos. 3–4, the circuit runs *abce* in leg *G* and leg *M*, exactly as in nos. 1–2.)

Now, it is understandable that Galvani should have regarded these reasons as convincing, even decisive. However, *logically speaking*, they were objectively inadequate to confirm his rigid "all or nothing" position. By repeating and varying his experiments on contractions—especially those with the animal arc between nerve and muscle (GO, 313, 317, 420; GOS, 433, 436, 487), those with

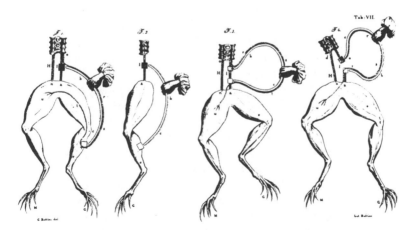

6.2 Plate 1 of Galvani's *Memorie sull'elettricità animale*.

the animal arc between nerve and nerve (GO, 343), and, above all, the one just quoted with the two nerves bridged by an arc consisting of the nerve itself—Galvani could rightly claim to have proved his own theory of animal electricity. But he could not additionally claim to have refuted Volta's theory—at least the general theory of contact electricity, which, after the third letter to Vassalli-Eandi (27 October 1795), was the only contender. Indeed, Galvani did not even bother to examine it.

This put Volta in a good position to counterattack. He did so in two anonymous letters to Aldini of April 1798, signed "Citizen N.N. of Como." To begin with, he reproached Galvani for his haste: "Now, this 'in other words,'[2] this denial that metals possess any virtue or power of unbalancing the electrical fluid, is what has hurt me—as it contradicts not only Volta's manifold arguments and proofs regarding Galvanism, but also other facts and direct experiments independent of galvanism" (VO, 1:525; VOS, 506). He then caught out Galvani, accusing him of overlooking his new theory—which by then was actually more than a year old. As a result, Volta argued, Galvani's latest replies were "forestalled and pretty much destroyed" (VO, 1:526; VOS, 508). Lastly, he reminded Galvani of the experiments on the contact electricity between two first-class conductors, between first- and second-class conductors, and also the electricity produced "by the contact between two plates, neither of them metallic, with plates of cardboard, wood, etc., sufficiently soaked, but not too much . . . with various liquids" (VO, 1:552). All that remained was for him to conclude that galvanism was untenable and his theory "demonstrated."

> Now, therefore, that it is [demonstrated], now that the matter is not only proven, but displayed before one's eyes by the above-described experiments with metallic and even non-metallic plates, which become greatly electrified and impart clear signs to the electrometer through mere mutual contact—what do you say, my dear Aldini, and what will Galvani himself say? My conclusion is that of most Physicists who have learned of these latest experiments, and who, even earlier—on the sole basis of the experiments on Galvanism multiplied and varied in so many ways by Volta—had embraced his opinion. I conclude, then, that Volta demonstrates his artificial and extrinsic electricity—an electricity excited, that is, by the mutual contact of any conductors, provided they are dissimilar, and especially if metallic; that he proves it by direct, very simple experiments, and in a certain sense enables us to touch it physically; that he discovers its type, that is, where it is *positive* or *plus*, and where it is *negative* or *minus*; and that he measures it with the electrometer. Galvani, instead, has failed to demonstrate the alleged animal electricity in his experiments (which in other respects are very beautiful, and surprising)—to wit, the electricity that, he claims, is moved internally by any vital force or organic function. Even less has he succeeded in making it perceptible by the electrometer, nor do I believe he will ever be able to. So with what arguments and proofs does he want to sustain it? And how will he still find supporters, if all those experiments, all the new attempts

hitherto performed in the area of Galvanism, all the new phenomena can be sufficiently accounted for by the principle discovered and now demonstrated with the utmost evidence by Volta—namely, the power, the action impelling the electrical fluid, that is deployed in every contact between dissimilar conductors? (VO, 1:555)

Shortly after, however, in a letter to Brugnatelli of 19 October 1798, Volta suggested a compromise hypothesis, as Galvani had done before him. Rehearsing opinions expressed earlier (in letters to van Marum, 30 August 1792; to Delfico, 13 April 1795; to Mocchetti, August 1795), he revived a widely held physiological theory that he had previously criticized as unscientific. The theory claimed that the human body contained an electrical fluid performing the function of the so-called animal spirits; the fluid was said to reside in the brain (the nerves' point of origin) and could therefore be set in motion by the will, even for short periods, giving rise to muscular motion.

Here then, after having done and written so much to disprove the alleged animal Electricity—the one that is excited by means of metals in severed limbs, etc.—I too admit a genuine animal Electricity in the Torpedo, the shaking Eel and other shock-imparting fishes. And I am also inclined to attribute a similar one, in the manner explained above, to all animals. This Electricity is all the more truly animal as it depends on the soul [*anima*] and therefore obeys the will, and has little or no actual extension outside its locus. Thus, as the will is absent in the Galvanic experiments on severed limbs, etc., the electrical fluid is moved no longer by an internal principle but by an external cause: the application of unlike conductors. These are true *motors*, as I have argued and continue to argue, having supplied the most direct proofs of this. To a certain extent, therefore, these external motors replace the internal mover, which in the natural state of life is the animal's will. If the Galvanists are pleased to reduce animal Electricity to these terms, I shall be very pleased to agree with them. If, however, they continue to reject this mode of conciliation, which I am happy to offer them; if they persist in pretending that Electricity is excited by sheer organic force—to wit, that the electrical fluid is prepared and works in the brain and nerves, accumulates in these, or on the inside of the muscles, becomes unbalanced in some manner, and is discharged because of this imbalance, immediately stimulating the muscles themselves; if, I repeat, they continue to claim that such electricity is produced by a purely organic mechanism, even in severed limbs or muscles, even in a small piece of muscle—whereas such muscles are made to contract with the device of the heterogeneous conducting arc; if they persist in denying—in the face of all the evidence I have produced—that in these cases, and all the analogous experiments of *Galvanism*, the force at work is an artificial electricity, that is, excited by external motors; in other words, if they do not subscribe to such a reconciliation plan, I may perhaps withdraw it too; that is, I may no longer even concede the existence of that other animal Electricity—determined and moved by the will in the whole, intact living being—except in the Torpedo and other electric animals, since ultimately it is a mere hypothesis, and it is only as such that I wanted to advance it. (VO, 1:561)

Volta's tone, one must admit, was frankly arrogant. But beyond the tone, let us look at the content of the proposal, repeated a few days later in a letter to Gian Pietro Frank (VE, 3:412–18). This was the second compromise hypothesis— after Galvani's—with which the contestants sought, at least apparently, to end the dispute to their reciprocal advantage. The hypotheses can be summed up as follows:

Galvani's compromise	*Volta's compromise*
There exists an animal electricity in nerve-muscle contact, and a metallic electricity in other cases.	There exists an animal electricity in living organisms and a contact electricity in other cases.

The two compromises were neither coextensive nor equivalent. They were not coextensive, because the cases in which they granted the existence of animal electricity did not coincide. The contractions produced by the mere nerve-muscle contact in a frog prepared in the usual manner were, according to Galvani, a case of animal electricity; for Volta, a case of contact electricity, because they occurred in severed limbs. More important, the hypotheses were not equivalent, because the animal electricity referred to by Volta was not—despite his wording—exactly the same as the one referred to by Galvani. In fact, the concept Volta offered in his compromise was an ordinary electricity existing *in the animal* "as in all other good conductors" (VE, 3:415) and put in circulation there by a stimulus; it was not an *animal* electricity, specific and exclusive to the living organism.[3]

We can now understand why Galvani's compromise, besides being rejected by Galvani himself, could not be accepted by Volta; and why, conversely, Volta's compromise could not be accepted by the Galvanists. For the latter was not a compromise at all, or at least it was only a verbal one. In reality, it was a rehearsal of Volta's unitary theory against the theory of the two electrical fluids—the animal and the common.

Galvani had been too hasty with his "all or nothing." But Volta was no less so. *By strict logic*, Galvani could not assert that his experiments were irrefutable evidence that the frog's electricity was *entirely* animal. After all, the animal's nerve and muscle might well be behaving like generic dissimilar conductors unbalanced through contact. On the other hand, Volta could not affirm that the electricity of the frog (even that of a freshly killed and prepared specimen) was *entirely* generated by the contact of two generic dissimilar conductors. In the animal, the general electromotive force might well have replaced or combined with a force specific to the nerves and muscles. Thus, while Volta accused Galvani of "excessive attachment to his preconceived opinion" (VO, 1:525n; VOS,

506n), it would have been easy and just as legitimate to level the same charge against Volta himself.

The fact remained that the two adversaries now moved on parallel tracks from which neither, in reality, had ever deviated by an iota: Galvani was intent on explaining contractions *iuxta principia biologica*, that is, on confirming the biological origin of the frog's current; Volta, instead, on explaining these and other phenomena *iuxta principia physica*, that is, on establishing the physical origin of all currents.

But, for one contestant, the end was approaching. On 14 May 1796, Napoleon Bonaparte entered Milan, and Volta was among those in charge of paying homage to him (VE, 3:321). On 19 June, the general reached Bologna. The French invasion caused a profound political and psychological upheaval. "The Catholic religion will always be an irreconcilable enemy of the republic," the Directory wrote in an instruction to Bonaparte. And so it was for the conscience of many. On 13 February 1797, Volta, together with three other faculty deans, protested in the name of religious values against the introduction of the French revolutionary calendar based on the ten-day week (VE, 3:350–53). He then formally "shirked" (VE, 3:465) the oath imposed by the Cisalpine Republic on public officeholders; he reached an accommodation with the new regime for which, like all the professors of the University of Pavia, he had to suffer a year's ostracism when the Austrians returned. Later, on 23 June 1800, when the regime had changed once more, Napoleon reinstated Volta as professor of experimental physics. Galvani, instead, remained faithful to his old values. He openly refused to take the oath, and was stripped of his post and salary on 20 April 1798. He survived—with dignity—just a few months, dying at only sixty-one on 4 December 1798. Galvani did not live to see the last act of the controversy. He would have emerged defeated—as many of his followers actually did—but not convinced. For henceforth it was not only experiments and instruments that were to speak for science—not even the truly "wonderful" instrument that Volta announced to the world just over a year later.

6.2 THE "COLUMNAR APPARATUS"

The twentieth of March 1800 is one of the epoch-making dates in the history of science. By terminating the debate that had gripped all of learned Europe, it ushered in a new age of physics. It is as if a new, hitherto inconceivable world had been discovered. On that day, Alessandro Volta wrote a long letter, in French, to Sir Joseph Banks, secretary of the Royal Society in London, announcing the invention of the pile. The document was immediately published with an English title—"On the Electricity excited by the mere Contact of conducting Substances of different Kinds"—in the *Philosophical Transactions* (vol. 90, pt. 2) and translated into English in *The Philosophical Magazine* (September

1800). It was written in haste, as amply evidenced by its very structure, fraught with redundancies and reconsiderations. In quoting the text as fully as it deserves, we must therefore organize it for an easier reading of the "striking results" it reported and of the goals it intended to achieve.[4]

"The principal of these results, which comprehends nearly all the rest, is the construction of an apparatus . . . which should have an inexhaustible charge, a perpetual action or impulse on the electric fluid." The apparatus was formed *exclusively* of a large number of "non-electric" bodies, "chosen from among those which are the best conductors, and therefore the most remote, as has hitherto been believed, from the electric nature" (VO, 1:566; Volta 1952, 42).

After this introductory announcement, the letter may be regarded as containing four parts. The first describes the "column" pile (fig. 6.3). Volta begins with the construction materials:

> I provide a few dozens of small round plates or disks of copper, brass, or rather silver, an inch in diameter more or less (pieces of coin for example) and an equal number of plates of tin, or, what is better, of zinc, nearly of the same size and figure. . . . I prepare also a pretty large number of circular pieces of pasteboard, or any other spongy matter capable of imbibing and retaining a great deal of water or moisture, with which they must be well impregnated in order to ensure success to the experiments. These circular pieces of pasteboard, which I shall call moistened disks, I make a little smaller than the plates of metal, in order that, when interposed

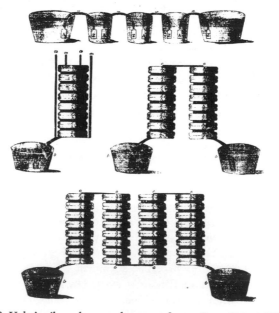

6.3 Volta's pile: column and crown of cups. From VO, 1:570.

between them, as I shall hereafter describe, they may not project beyond them. (VO, 1:566; Volta 1952, 42)

The construction follows:

I place then horizontally, on a table or any other stand, one of the metallic pieces, for example one of silver, and over the first I adapt one of zinc; on the second I place one of the moistened disks, then another plate of silver followed immediately by another of zinc, over which I place one of the moistened disks. In this manner I continue coupling a plate of silver with one of zinc, and always in the same order, that is to say, the silver below and the zinc above it, or *vice versa*, according as I have begun, and interpose between each of these couples a moistened disk. I continue to form, of several of these stories, a column as high as possible without any danger of its falling. (VO, 1:567; Volta 1952, 42)

Then come the first effects:

But, if it contain about twenty of these stories or couples of metal, it will be capable not only of emitting signs of electricity by Cavallo's electrometer, assisted by a condenser, beyond ten or fifteen degrees, and of charging this condenser by mere contact so as to make it emit a spark, &c., but of giving to the fingers with which its extremities (the bottom and top of the column) have been touched several small shocks, more or less frequent, according as the touching has been repeated. Each of these shocks has a perfect resemblance to that slight shock experienced from a Leyden flask weakly charged, or a battery still more weakly charged, or a torpedo in an exceedingly languishing state, which imitates still better the effects of my apparatus by the series of repeated shocks which it can continually communicate. (VO, 1:567; Volta 1952, 42–43)

Finally, a few precautions:

To obtain such slight shocks from this apparatus which I have described, and which is still too small for great effects, it is necessary that the fingers, with which the two extremities are to be touched at the same time, should be dipped in water, so that the skin, which otherwise is not a good conductor, may be well moistened. To succeed with more certainty, and receive stronger shocks, a communication must be made, by means of a metallic plate sufficiently large, or a large metallic wire, between the bottom of the column (that is to say, the lower piece of metal,) and water contained in a bason or large cup, in which one, two, or three fingers, or the whole hand is to be immersed, while you touch the top or upper extremity (the uppermost or one of the uppermost plates of the column) with the clean extremity of another metallic plate held in the other hand, which must be very moist, and embrace a large surface of the plate held very fast. . . .

I still suppose that all the necessary attention has been employed in the construction of the column, and that each pair or couple of metallic pieces, resulting from a plate of silver applied over one of zinc, is in communication with the following couple by a sufficient stratum of moisture, consisting of salt water rather than common water, or by a piece of pasteboard, skin, or any thing of the same kind well impregnated with this salt water. The disk must not be too small, and its surface must adhere closely to those of the metallic plates between which it is placed. (VO, 1:567–68; Volta 1952, 43)

The second part of the letter contains the description of a variant of the columnar apparatus, the *couronne de tasses* pile ("crown of cups," rendered in the 1800 translation as "chain of cups"). Here again, Volta begins by describing the construction of the apparatus (fig. 6.3):

I dispose, therefore, a row of several basons or cups of any matter whatever, except metal, such as wood, shell, earth or rather glass (small tumblers or drinking glasses are the most convenient), half filled with pure water, or rather salt water or ley: they are made all to communicate by forming them into a sort of chain, by means of so many metallic arcs, one arm of which, [*Aa*], or only the extremity [*A*], immersed in one of the tumblers, is of copper or brass, or rather of copper plated with silver; and the other, [*Z*], immersed into the next tumbler, is of tin, or rather of zinc. (VO, 1:569–70; Volta 1952, 43–44)

After a few instructions for the best use, Volta goes on to relate a few experiments "which are no less instructive than amusing":

Let three twenties of these tumblers be ranged, and connected with each other by metallic arcs, but in such a manner, that, for the first twenty, these arcs shall be turned in the same direction; for example, the arm of silver turned to the left, and the arm of zinc to the right; and for the second twenty in a contrary direction, that is to say, the zinc to the left, and the silver to the right: in the last place, for the third twenty, the silver to the left, as is the case in regard to the first. When every thing is thus arranged, immerse one finger in the water of the first tumbler, and, with the plate grasped in the other hand, as above directed, touch the first metallic arc (that which joins the first tumbler to the second), then the other arc which joins the second and third tumbler, and so on, in succession, till you have touched them all. If the water be very salt and luke-warm, and the skin of the hands well moistened and softened, you will already begin to feel a slight shock in the finger when you have touched the fourth or fifth arc (I have experienced it sometimes very distinctly by touching the third), and by successively proceeding to the sixth and the seventh, &c., the shocks will gradually increase in force to the twentieth arc, that is to say, to the last one of those turned in the same direction; but by proceeding onwards to the 21st, 22d, 23d, or 1st, 2d, 3d, of the second twenty, in which they are all turned in a contrary direction, the shocks will each time become weaker, so that at the 36th or 37th, they will be imperceptible, and be entirely null at the 40th, beyond which (and beginning the third twenty, opposed to the second and analogous to the first,) the shocks will be imperceptible to the 44th or 45th arc; but they will begin to become sensible, and to increase gradually, in proportion as you advance to the

60th, where they will have attained the same force as that of the 20th arc. (VO, 1:572; Volta 1952, 44)

There follows a theoretical consequence of these experiments, added almost as an aside—but an important aside, because it strikes at the heart of the animal-electricity theory, precisely at the point on which animal electricity enjoyed a universal consensus:

> From these experiments one might believe, that when the torpedo wishes to communicate a shock to the arms of a man or to animals which touch it, or which approach its body under the water (which shock is much weaker than what the fish can give out of the water), it has nothing to do but to bring together some of the parts of its electric organ in that place, where, by some interval, the communication is interrupted, to remove the interruptions from between the columns of which the said organ is formed, or from between its membranes in the form of thin disks, which lie one above the other from the bottom to the summit of each column: it has, I say, nothing to do but to remove these interruptions in one or more places, and to produce there the requisite contact, either by compressing these columns, or by making some moisture to flow in between the pellicles or diaphragms which have been separated, &c. This is what may be, and what I really conclude to be, the task of the torpedo when it gives a shock; for all the rest, the impulse and movement communicated to the electric fluid, is only a necessary effect of its singular organ, formed, as is seen, of a very numerous series of conductors, which I have every reason to believe sufficiently different from each other to be *exciters* of the electric fluid by their mutual contacts; and to suppose them ranged in a manner proper for impelling that fluid with a sufficient force from top to bottom, or from the bottom to the top, and for determining a current capable of producing the shock, &c. as soon and as often as all the necessary contacts and communications take place. (VO, 1:573–74; Volta 1952, 45)

In the third part, Volta returns to the "columnar apparatus," giving instructions on how to build the pile, divide the column, and keep the pile running efficiently.

In the fourth and final part, Volta resumes his discussion of effects other than commotions:

> The current of the electric fluid, impelled and excited by such a number and variety of different conductors, silver, zinc, and water, disposed alternately in the manner above described, excites not only contractions and spasms in the muscles, convulsions more or less violent in the limbs through which it passes in its course; but it irritates also the organs of taste, sight, hearing, and feeling, properly so called, and produces in them sensations peculiar to each. (VO, 1:576; Volta 1952, 46)

Here too, Volta breaks his detailed analysis of the effects to slip in a theoretical conclusion almost as an aside:

What proof more evident of the continuation of the electric current as long as the communication of the conductors forming the circle is continued?—and that such a current is only suspended by interrupting that communication? This endless circulation of the electric fluid (this *perpetual motion*) may appear paradoxical and even inexplicable, but it is no less true and real; and you feel it, as I may say, with your hands. Another evident proof may be drawn from this circumstance, that in such experiments you often experience, at the moment when the circle is suddenly interrupted, a shock, a pricking, an agitation, according to circumstances, in the same manner as at the moment when it is completed; with this only difference, that these sensations, occasioned by a kind of reflux of the electric fluid, or by the shock which arises from the sudden suspension of its current are of less strength. But I have no need, and this is not the place to bring forward proofs of such endless circulation of the electric fluid in a circle of conductors, where there are some, which, by being of a different kind, perform, by their mutual contact, the office of exciters or *movers*: this proposition, which I advanced in my first researches and discoveries on the subject of galvanism, and always maintained by supporting them with new facts and experiments, will, I hope, meet with no opposers. (VO, 1:576–77; Volta 1952, 46)

Volta concludes his letter with a similar argument:

To what electricity then, or to what instrument ought the organ of the torpedo or electric eel, &c. to be compared? To that which I have constructed according to the new principle of electricity, discovered by me some years ago, and which my successive experiments, particularly those with which I am at present engaged, have so well confirmed, *viz.* that conductors are also, in certain cases, exciters of electricity in the case of the mutual contact of those of different kinds, &c. in that apparatus which I have named the *artificial electric organ*, and which being at the bottom the same as the natural organ of the torpedo, resembles it also in its form, as I have advanced. (VO, 1:534; Volta 1952, 49)

Had Volta scored a total victory? Let us examine the question in greater detail.

6.3 THE PILE AND THE THEORY OF CONTACT ELECTRICITY

Volta's letter has often been characterized as more remarkable for what it does *not* say than for what it says (see Figuier 1868, 624; Sirol 1939, 63; Polvani 1942, 347; Mauro 1969, 149). We know that the letter was in fact "an excerpt from a long paper . . . consisting of various fragments that were juxtaposed at random with gaps left" (VO, 2:15); that the letter had a sequel; that Volta promised himself to return to the subject. But all this does not suffice to explain its content. The letter contains silences, misunderstandings, and overstatements that cannot be attributed to haste or to the circumstances of its writing.

First, the silences. There is not a single word in the entire text on the chemical effects of the pile. There is nothing on oxidation or on the formation of gas bubbles in the crown of cups. It is impossible that these phenomena never occurred in the presence of the chemist Volta. It is just as hard to believe that this highly attentive and scrupulous experimenter failed to notice them. Why then did he not mention and analyze them?[5]

The misunderstandings contained in the letter give us a first indication. Volta correctly observed and remarked that the column pile "does not long continue in a good state" (VO, 1:575; Volta 1952, 45) but attributed this to fact that "the moistened disks become dry in one or two days." To overcome this problem, he said that "they must again be moistened" and suggested this "may be done without taking to pieces the whole apparatus, by immersing the columns, completely formed, in water." This was a misunderstanding but—worse still—an inconsistency, because the fading also occurred in the crown of cups, where one could hardly speak of a dry-out effect. Here again, Volta failed to note the phenomenon. Why?

Nor was this the only case. The other misunderstanding concerned the function of the moist layers. In the letter to Banks, they appeared as simple conductors—some less efficient, such as plain water; others more efficient, such as salt water or lye. Volta later reiterated the concept. "The humid strata employed in these complicated apparatus," he wrote on 10 October 1801 in a letter to Jean-Claude de La Métheric, "are applied therefore for no other purpose than to effect a mutual communication between all the metallic pairs" (VO, 2:40; English translation from Volta 1802, 138; see VO, 2:162, 163, 289); as for the saline liquids, they "merely facilitate the passage, and leave a freer course for the electric fluid, being much better conductors than simple water, as several other experiments prove" (VO, 2:41; Volta 1802, 139; see VO, 4:228; VOS, 587). Nor did Volta change his mind or demonstrate any suspicion when he observed "in passing" in the letter to Banks that "ley and other alkaline liquors are preferable when one of the metals to be immersed is tin: salt water is preferable when it is zinc" (VO, 1:571; Volta 1952, 44, where "j'observerai ici, en passant" in the French original becomes simply "I shall here observe")—that is, when he was faced with clearly chemical phenomena. Why?

The reason lies precisely in the explicitly chemical quality of these phenomena—a quality Volta did everything to exorcise—and in the hypothesis they inevitably suggested. In his letter to Banks, Volta wanted not only to confine himself to a description of the pile and its effects; he wanted above all to offer an explanation of the apparatus consistent with his theory of contact electricity. His description was therefore governed and conditioned by the theory he supported; and since this theory "gives all to the metals and nothing to the fluids," as Nicholson commented critically (Nicholson 1802, 143; VO, 2:154n), Volta misinterpreted the chemical phenomena, failed to note them, or—if he did observe them—left them out of his report.

The overstatements in the letter to Banks are in this respect as revealing as the omissions. The lengthy fourth part of the document is entirely taken up with physiological experiments and the strongest or most "paradoxical" expressions are introduced for the purpose of presenting the pile as a "perpetual" electromotor. Elsewhere, Volta referred to the device as an "artificial electric organ" or by the nonce-word *elettrotomeno perpetuo* (apparently from the Greek μένος, "force"—a "perpetual electrical force"; VO, 2:139n). Volta saw the pile as a reproduction of the system used by the torpedo. Indeed, for Volta, the torpedo's organ *was* a pile. In this light, Volta's entire aim becomes unambiguously clear: he was pursuing his old strategy—already applied with success in the case of the electrophore and condenser—of using an instrument to "confirm," as he wrote at the end of the letter to Banks, or "better yet, to establish," as he said shortly after (VO, 2:16), a theoretical principle. This strategy now reached its peak. The pile's operation was, for Volta, the clearest sign of the truth of his contact theory. And since the theory was what *had* to be established, everthing incompatible with it was either not observed, or, if observed, was passed over in silence, or, if mentioned, was reinterpreted ad hoc. In biographical terms, this may perhaps be explained by the researcher's stubbornness;[6] in psychological terms, by the illusionistic effect of expectations. At all events, the truism remains—but it is hardly trivial—that they who seek shall find, especially if they are strongly committed to the object of their quest.

As earlier, Volta was successful. More specifically, his purpose was twofold: to show that an electromotive force was generated by the contact between two dissimilar conductors, especially metallic ones; and to show that the electrical fluid thus generated was the same as the galvanic fluid. The pile won him consensus on both points. On the first point, however, the contact theory had a checkered career from the outset. It encountered immediate opposition from Nicolas Gautherot in France; William Nicholson, Humphry Davy, and William Hyde Wollaston in England; and Georg Friedrich von Parrot in Russia, while in Italy the theory of the chemical origin of the electromotive force of metals had already been advanced by Giovanni Fabroni.[7] But Volta's contact theory soon gained the support of scientists in most countries, who followed the influential examples of Biot, Pfaff, and van Marum. It was only in the 1830s and 1840s, after the work of Arthur-Auguste de La Rive and Michael Faraday, that the scientific community changed its mind on the subject—this time for good.

The second point—the identity of the electrical and galvanic fluids—was also initially contested. On 7, 12, and 22 November 1801, Volta read before the Institut de France, in Bonaparte's presence, the "Memoria sull'identità del fluido elettrico col fluido galvanico." On 2 December, Biot read a "Rapport sur les expériences du citoyen Volta," praising the single-fluid theory (VO, 2:114–15).

The same day, at the end of the session, Biot, at Bonaparte's behest, presented Volta with the medal of the Institut. Volta, then fifty-six and at the peak of his career, was the most famous physicist in Europe. Admittedly, he met with resistance from scientists such as Aldini (1802, pt. 3; 1803; 1804), Vassalli-Eandi (1803a, 1803b) and Deluc (1804), who in various ways defended the unlikely existence of a galvanic fluid displaying somewhat different properties from common electrical fluid. Although none of these opponents produced new or better arguments, Volta decided to reexamine the entire issue. In 1805 he drafted a long paper that he communicated to his pupil Giuseppe Baronio for submission to the contest sponsored by the Società Italiana delle Scienze of Modena. After Baronio's death in 1814, the paper was published by Pietro Configliachi under the title L'identità del fluido elettrico col così detto fluido galvanico vittoriosamente dimostrata con nuove esperienze ed osservazioni.

This was truly "Volta's last triumph" (Luigi Magrini, quoted by Alessandro Volta, Jr., in VO, 2:207). And while it did not impose "perpetual silence on the various sects of Galvanists," as Configliachi rhetorically proclaimed (VO, 2:201), it is a fact that for thirty years after the discovery of the pile virtually no one ever ventured to speak of animal electricity again. The first to break the silence—albeit cautiously, as this was still a burning issue—was Leopoldo Nobili. The next to intervene, with resolve, was Carlo Matteucci. Historically speaking, therefore, Volta won this match too, at least for a while.

Concerning objective merit, however, the picture is quite different. As in the 1796–97 experiments, Volta could not argue that the pile simply "dispelled beyond any doubt my new principle of electricity" (VO, 2:32). Volta claimed to have proved that the right combination of metals and moist conductors would produce electricity, and that this electricity was identical to the galvanic one. But this did not imply that the electricity in animals—for example in torpedoes, not to speak of the frog—was not specific and peculiar to them. To reach Volta's conclusion from his premises required a logical leap that could be taken only by introducing another premise—which, as we shall see, constituted the true stake in the controversy.

Volta made the leap when he transferred the pile's operating mode to animals. On this point he even revised his earlier views. While in the past he had always argued that a true animal electricity surely existed in electric fish, in the letter to Banks he held that not even the electricity of fish could properly be called animal. He later returned to the subject and wrote:

> Although metals and other first-class conductors are in general, even as motors, more effective than moist or second-class conductors, it is possible even with the latter alone—provided they are suitably different from one another—to form fairly powerful Electro-motors, even with only animal or vegetable substances.
>
> The so-called electrical organs of the Torpedo and of the other fish that possess the wonderful virtue of imparting shock seem to be of this kind—indeed, they can

certainly be said to be so. Thus, not even in this case is it proper to speak of *animal electricity*, in the sense of being produced or moved by a truly vital or organic action, as such is not the case here. Rather, it is a simple physical, not physiological, phenomenon—a direct effect of the Electro-motive apparatus contained in the fish. This apparatus is similar to the artificial Electro-motors and, like these, it acts on its own force, by virtue of its construction, that is, of the mutual contacts between unlike conductors, etc. (VO, 2:296; VOS, 640–41)

This is yet another illustration of the "all or nothing" logic shared by Volta and Galvani. And once again, as with Galvani, we can observe the distortions it engendered. Two points concerning Volta's attitude are worth highlighting here.

The first relates to the incompleteness of the pile-torpedo analogy. The general theory of contact electricity ruled out the possibility of generating a continuous electrical current with conductors of a single class. The analogy, instead, made that assumption. Volta was then faced with a choice: (1) Either "we must say that the relationship and the regular gradation indicated with respect to the motive powers of the electrical fluid [that is, the law of contacts] are not observed in second-class conductors—not even among themselves—or at least do not apply to all of them, but that things proceed with other laws and in a different mode, at least as regards some conductors." This would require making an exception to the theoretical law of contact. (2) Or "we may perhaps have to subdivide the second class and recognize a third class, in which the conductors are related to one another in their motive property—like the conductors within each of the other two classes—but are not linked in this respect with the second-class conductors." This alternative required an exception to the empirical law of conductors (VO, 2:62–63; VOS, 577). Since the first exception included, at least in principle, the possibility of an animal electricity, Volta preferred the second. But this exception meant that the torpedo or animal could no longer be termed a common pile. It was a pile of a special kind: an animal pile that could in no way be reduced to an artificial pile and explained in terms of the latter's operation.

The other point to note concerns the equivalence of the general theory of contact electricity and Galvani's theory as regards the electricity ascertained in animals. The special theory of contact electricity held animal organs to be "merely passive" (VO, 1:146); the general theory too regarded them as "merely passive" (VO, 1:523; VOS, 503), while arguing that even dissimilar conductors belonging solely to the second class were endowed with an electromotive force. But this complicated matters, for the second-class conductors whose contact had the power to set the fluid in motion in Valli and Galvani's experiments (contractions without metals) were precisely the animal organs—that is, nerve and muscle. As a result, on the basis of the general theory of contact electricity itself, nerve and muscle were not passive but active. Hence, the general theory of contact was equivalent to the theory of animal electricity, because the core of the latter was that nerve and muscle, being naturally imbalanced, were *electro-motors*. The two theories were thus no longer at odds with each other. Or rather: the general theory of contact clashed with the theory of animal electricity only if nerve and muscle were regarded as *animal organs* and not, in any way, as generic *moist conductors*.

"Who can fail to see that we're both saying the same thing, albeit with different words?" Galvani had written when tempted to compromise (GO, 322; GOS, 438). However, the controversy hinged not on *words* but on *concepts* or, more accurately, *presuppositions*. A physical presupposition allowed Volta to conclude from his experiments that first the torpedo, then the frog, then all animals in general were piles; a biological presupposition enabled Galvani to conclude from the same experiments that the frog was an organic Leyden jar. There lay the true kernel of the dispute. Let us try to understand its logic better.

6.4 THE AMBIGUOUS FROG

To clarify the function and effects of presuppositions in scientific research, recent philosophers of science have introduced various technical concepts such as "paradigms" (Kuhn 1962) or "natural interpretations" (Feyerabend 1965; 1975). Elsewhere as well as in this work (§3.1) I have used the notion of *interpretative theories* as distinct from *explanatory theories*.

An explanatory theory is an explanation in theoretical terms of laws and phenomena, or a solution to certain empirical problems. In our case, we can regard as explanatory theories: (1) Galvani's theory of an electrical fluid distributed in a naturally unbalanced manner among the different parts of animal bodies; (2) Volta's theory of an artificial imbalance produced by the electromotive force either of metals (special theory) or of generic dissimilar conductors (general theory) brought into contact.

An interpretative theory is a broader theoretical unit. Its purpose is to prescribe and assign a certain structure to the world or to parts of the world; it indicates and classifies the types of entities that compose the world. These entities—or ontology—provide the basic concepts and the vocabulary on which to build explanatory theories. Here, we have two interpretative theories: (1) Galvani's notion that the frog contractions belong to the class of biological phenomena; (2) Volta's notion that the contractions belong to the class of physical phenomena. In general, as our example shows, interpretative theories are hidden—in other words, tacitly admitted.

Two main points seem to distinguish interpretative theories—as understood here—from both Feyerabend's natural interpretations and Kuhn's paradigms. First, like Feyerabend's natural interpretations, interpretative theories have to be conceived as constitutive of their domain or the field of observations they refer to, but not as natural—except in the sense that, once adopted, they become a habit of the inquiring mind. Interpretative theories can thus be brought to light and discussed critically. Such a discussion, although impossible to base "crucially" on facts, cannot be likened to a dialogue between two deaf persons, to be solved by an act of force or by a form of illusion. Rather, it is a real debate in which arguments and counterarguments face each other, and advantages and disadvantages are weighed.

Second, interpretative theories as defined here, at least Galvani and Volta's, function as gestalten with respect to the experiments in their domain, giving rise to *observationally equivalent* conceptual systems. This point is not completely clear in Kuhn. Although he compares paradigms to gestalten, he speaks of gains and losses in paradigm change, apparently implying that a later paradigm does not "gestaltically" restructure the *entire* domain of an earlier one.[8]

To say that two explanatory theories are observationally equivalent means (1) that they produce different, mutually translatable organizations and structurings of the same empirical domain; (2) that, as a result, their respective objects are incommensurable; (3) that, because of this, the two theories are undecidable on the basis of the structured data and of any datum from the domain.

Let us try to understand whether this is actually the case here. We begin with Volta's general theory of contact electricity. As we know, this theory holds that an electrical current is produced whenever at least *three* conducting bodies are put into contact. Of these conductors, two compose the arc and generate the electromotive force, and one or more form the "full" arc, that is, they close the circuit. In the special case of the frog, the arc is composed either of two metals, or of a metal and a liquid, or of the nerve and muscle; the arc is "closed" respectively by the frog (nerve + muscle), the frog again, and the liquid. Thus the observational domain of Volta's theory consists of a compound heterogeneous arc and a terminator—itself dissimilar with respect to the two conductors forming the arc.

But this very same set of elements also constitutes the observational domain of the competing theory. Granted, Galvani repeatedly insisted that, in his first experiments on frog-muscle contraction, the arc was a homogeneous copper device. In the same vein, Aldini emphasized that the mercury arc was homogeneous. And all the Galvanists maintained that the animal arc consisted of homogeneous organic tissue. Yet, if we look more closely, we find that even Galvani—when he spelled out his theory in detail rather than simply trying to refute Volta's—came to conclude that muscular contractions occurred only with a compound heterogeneous arc.

To begin with, let us look at the composition of the natural arc—the one that in every living organism must establish communication between the internal and external parts of the muscle fiber. In the *Trattato*, where Galvani raises the problem, he clearly states that "the substance of the nerve does not suffice to form the complete arc" (GO, 242; GOS, 407). The reason is obvious: such a substance "would conduct the electricity from the inner part of the muscular fibers . . . to the nerve, but would not actually return it to that muscle." The surplus electricity in the muscle, collected by the nerve, must therefore be diffused through "extrinsic moist parts": "Thus, for contractions to occur, not only must certain internal parts carry electricity from the muscle to the nerve, but other, external parts must carry it back from the nerve to the muscle." The process therefore requires a "natural internal arc" and a "natural external arc." The first "one can conjecture as being formed of the pure medullary substance of the nerve"; the second must be assumed to consist of "moisture, the sheaths of the nerve itself, and the humors contained in these sheaths." Consequently, Galvani concluded, "if we subscribe to the Leyden-jar hypothesis as the most plausible and—of all those so far put forward—the one that best explains the phenomena, there would be a dissimilar arc consisting of two parts: the first, of nervous substance; the second, of membranes and humor. One end of the arc would communicate with the inner part of the muscular fibers, the other end with the outer part" (GO, 245; GOS, 408).

But this applied not only to the natural arc. In the *Memoria quarta* to Spallanzani, Galvani concluded that even the apparently homogeneous arc—metallic or not—was in fact a compound, heterogeneous arc without which no contractions occur. He raised the question—an anomaly for his theory—of why the heterogeneity makes "the contractions so much sharper" (GO, 368; GOS, 465) and solved it by arguing that "it can do this only by increasing the velocity or amount of the electricity forming the said current." Both effects were due "first, to the passage from a metal less conductive of animal electricity to another more conductive; second, to the passage of a more conductive to a less conductive metal; third, and lastly, to the contact between dissimilar metals themselves" (GO, 370; GOS, 466). In the first case, according to Galvani, the fluid would accelerate; in the second, it would accumulate; but in both cases one could observe that

> the contractions will be all the greater if the differences in conductivity between the metals are greater; and that the metals will produce no contractions whatsoever if there are no differences between them. Indeed, this is what we ordinarily observe in these experiments, in which the muscular contractions tend to be excited all the more violently as the above-mentioned difference is greater, and often do not take place when the metals are homogeneous. (GO, 371; GOS, 467)

The third case Galvani interpreted as follows. At the point of contact between dissimilar metals, there always remains a layer of air that hinders the flow of the electrical fluid. The thinner the layer, the greater the flow; and since the layer disappears completely if replaced by humidity, "even a minimal humidity at the point of contact between the arc and the homogeneous armatures will consis-

6.4 The three standard situations in which muscular contractions are obtained: (a) with a bimetallic arc; (b) with a single-metal arc; (c) with an animal arc. To the right: the corresponding arrangement of conductor piles in Volta's system.

tently substitute for the dissimilarity of the armatures" (GO, 376; GOS, 469–70). Thus, even a homogeneous metal arc is in reality a dissimilar arc composed of two parts: the metal and the humidity. Various experiments reviewed by Galvani proved this hypothesis, of which he said that "everyone sees it can be fairly conveniently applied to the dissimilarity of other bodies used as armatures" (GO, 378; GOS, 471) and therefore to the case where the conducting arc is composed of nonmetallic bodies and animal substances.

Galvani admitted one exception to the dissimilarity rule: the case of the nerve folded back on the muscle, where the "contact of the arc formed by the nerve alone is enough to excite the muscular movements" (GO, 379). However, if—as he claims here—this is the case "of the natural arc alone," then (1) as already stated in the *Trattato*, we are dealing here in fact with a dissimilar arc, composed of nerve and humidity; (2) it is hard to deny the presence of a moist layer in the point of contact between the muscle and the severed nerve; (3) in any event, this case is exceptional, and the standard situation calls for a dissimilar arc instead.

Thus, for Galvani no less than Volta, the contractions occur when *three* dissimilar conducting bodies are brought into contact, of which two form the arc, and one completes the arc. The conclusion is therefore that the relevant observational domain—the experimental situation that Volta and Galvani's theories aim to explain—was identical and shared by both scientists (fig. 6.4). In particular, Galvani and Volta recognized that the frog's muscular contractions are triggered by the contact of three dissimilar bodies; they recognized that the greater the dissimilarity, the larger the contractions; and they recognized that the greater the difference between the dissimilar conductors in contact—that is, their distance on the conductivity scale—the more vigorous the contractions.

From this common domain, however, the two positions diverged. To begin with, the *explanations* differed. Galvani attributed the contractions to a natural imbalance, inherent in nerve and muscle, that generated a current of fluid from the internal muscle to the external muscle via the nerve. Volta explained the contractions by arguing that, because of the contact between the two conductors composing the arc, the fluid became imbalanced, generating a current toward the metal or moist body of lower rank in the scale of first- or second-class conductors.

These two different explanations are also linked to two different *descriptions*. Galvani described the observational domain as composed of a nerve, a muscle, and one or more conductors, while Volta described it as consisting of three or more conductors. These unlike descriptions in turn depend on two different *perceptions*. To put it concisely, we could say that Galvani looked at the observational domain and saw—or saw it as—a condenser (Leyden jar); Volta looked at the same field and saw—or saw it as—an electromotor (pile).[9]

Thus the two theories are *different structurings of the same observational domain*; and their respective objects—that is, the significant entities produced by the theories from the same data—are dissimilar and incompatible. However, this could simply mean that the theories interpreted the same phenomena in different ways. On this basis alone, therefore, we cannot conclude that the theory of animal electricity and the general theory of contact electricity are observationally equivalent. One crucial point remains to be proved, namely, that the theories are undecidable—whether on the basis of the organized observational data or even on the basis of any datum that could possibly be acquired in their domain.

To clarify this point, let us take the third standard situation (fig. 6.4c). In order to prove the animal-electricity theory, Galvani performed the crucial experiment of putting a nerve directly in contact with a muscle. Observing that this gave rise to contractions, he concluded that nerve and muscle were naturally unbalanced and that the frog was a condenser. Meanwhile, to prove the general theory of contact electricity, Volta performed the equally crucial experiment of bringing into contact two unlike metals, a metal and a moist body, and two unlike moist bodies. Observing that in all three cases the two bodies in contact displayed contrary electrical signs, he concluded that two dissimilar conductors were capable of displacing the fluid and that the frog was an electromotor.

Both experiments are indeed crucial, but each is so on one condition. Galvani's is crucial provided one regards nerve and muscle precisely as nerve and muscle, that is, as two *animal organs*. Thus, as the imbalance lies between animal organs, the unbalanced electricity will be an animal electricity, and the cause of the imbalance will be intrinsic and natural. Volta's experiment, by contrast, is crucial provided nerve and muscle are viewed as *generic moist conductors*. Thus, as the imbalance lies between two generic conductors, the unbalanced electricity will be the electricity common to all bodies and the cause of the imbalance will be extrinsic and artificial. In other words, Galvani treats nerve and muscle as specific biological organs and therefore attributes the electricity they display to a biological cause; Volta treats nerve and muscle as belonging to the moist-conductor class and therefore ascribes the electricity they exhibit to a physical cause. To put it differently, for Galvani, "nerve" and "muscle" are proper names of biological entities, while for Volta they are common names of physical entities.

But what are these two different ways of looking at the same observational data? They are what we have called "interpretative theories," in the sense of two views that interpret or specify the basic phenomena—two ways of typifying or classifying or distributing. Indeed, they are two gestalten, which predefine the domain to which the phenomena belong—and the prior definition of the domain decides what constitutes a crucial proof. Thus Galvani's explanatory theory—the theory of animal electricity—is proved "crucially" provided we accept his interpretative theory, the electrobiological gestalt. Similarly, Volta's explanatory theory—the general theory of contact electricity—is proved "crucially" provided we accept his interpretative theory, the electrophysical gestalt. But these conditions are precisely what constitutes the stake in the dispute—the true bone of contention. And since neither Galvani nor Volta accepted his adversary's gestalt, there could be no genuinely crucial experiment to decide between the two theories. True, the theories conflicted, but there are no predictions made by one and denied by the other. Hence they are observationally equivalent, because each translates into its own language the facts, statements, and problems that appear in the language of the other.[10]

We may compare this situation of conflict with the contrasting "readings" or perceptions of the well-known ambiguous or multistable shapes such as Joseph Jastrow's duck/rabbit, E. Rubin's front/back cube, and E. G. Boring's mother-in-law/girl (fig 6.5). In each instance, the same perceptive material is naturally self-organized into different forms; and in each instance, the natural organization depends on the interpretative theory or gestalt we adopt—for the most part in an unspoken, unwitting, spontaneous manner.

6.5 Ambiguous shapes: (a) Duck or rabbit? (b) Organic condenser or physical electromotor?

This explains why Galvani and Volta alike reached a rigid, exclusive "all or nothing" position. Under the impulse of different interpretative theories, the two explanatory theories clashed without any margin for compromise, exactly as two ambiguous figures exclude each other. However, there was one significant difference: while the ambiguous figures contain no details incompatible with the preferred reading, the equivalent explanatory theories of Galvani and Volta involve distortions to force certain anomalies into the chosen interpretative theory. Galvani, for example, had to introduce special explanations to enable his framework to accommodate the increase of contractions with dissimilar arcs and the occurrence of contractions with the arc on the nerve alone. Volta, for his part, had to introduce special explanations to account for the current in the torpedo, an animal apparently composed of only second-class conductors.

Given these gestaltic effects, we must now ask ourselves on what basis the decision could be—and was—made in favor of one of the theories. As they are observationally equivalent, the choice cannot be made on an experimental basis.[11] Is it irrational then? Fearing this, Antonio Cima followed Spallanzani's line of argument and observed: "If a galvanic frog can display galvanic phenomena without any external device, reason dictates the statement that the frog inherently contains the sufficient cause of such phenomena. If the metal arc causes the frog to contract more violently or to contract when the mere contact between its parts no longer suffices for the purpose, we must say that the metal arc contains a new cause capable of shaking the frog" (1846, 139).[12]

But, as events showed, the "reason" cited by Cima or the "sound logic" invoked by Spallanzani—more properly, inductive logic—did not suffice to settle a controversy of this kind, not even to call it a draw. While such "reason" or "logic" prepares the decision, it cannot make the decision single-handedly. It is not hard to see why. If the contractions are obtained without artificial means, is their cause not intrinsic? Yes, it is intrinsic, answers Galvani's "reason." Yes, it is intrinsic, echoes Volta's "reason." But then Galvani, interpreting the phenomenon in the light of his electrobiological gestalt, specifies that Yes, the cause is intrinsic in the sense that distinct parts of the animal are electrically unbalanced; while Volta, guided by his electrophysical gestalt, states that No, the cause is intrinsic in the sense that distinct parts of the animal, when brought in contact, unbalance the electrical fluid in the same way as two generic moist bodies.

Thus, if inductive logic leads us to affirm that a transfer of fluid occurs between animal organs in the standard experimental situations, it does not also allow us to conclude whether the cause of the transfer is biological or physical. At the brink of such a conclusion—when the gestalt that assigns the phenomena to their proper sphere comes into play—inductive logic stops. But inductive logic does not exhaust reason, and there are no grounds for supposing that the lack of inductive proofs is tantamount to a lack of reasons. In cases like these, we are *beyond inductive logic*, but it would be a Cartesian non sequitur to conclude that we are *beyond reason*.

Where, then, are we? More precisely, on what *other* rational basis than the empirical grounds considered by inductive logic can we express a preference between two observationally equivalent theories such as Galvani and Volta's? And does such a basis really exist? In other words, is there a method for settling our controversy in a rational manner? The problem is twofold. First, it is historical: Was such a method actually at work at the time the preference was expressed? Second, it is philosophical: Can it work in principle, however things turned out in practice? Let us examine both aspects.

6.5 FROGS AGAINST METHOD[13]

The detailed story of the fate of Volta's theory remains to be written.[14] However, at least two factors can confidently be said to have contributed to its victory over Galvani's. We have already seen the first: Volta managed to pass off an instrument, the pile, as something it was not—the living confutation of Galvani's theory. The second factor was a mistake on the part of Volta's opponents. When you are engaged in a dispute, a wrong move may prove fatal. Greatly surprised by the pile, Galvani's followers tried to confront Volta's ideas by introducing a hypothesis that postulated the existence of either (1) two distinct electrical fluids, the common and the galvanic, or (2) a single fluid that, in animals, underwent some change and acquired new, special properties.[15] As the hypothesis proved untenable, its collapse automatically gave the edge to Volta's ideas.

There are many other factors we could mention, such as the suspicion with which most scientists viewed the galvanic experiments, many of them no more than funfair stunts put on by tramps and charlatans. But all this is still not enough to explain Volta's victory, especially if we are looking for an internalist, methodological explanation. If in a dispute a mistake by one camp turns into an advantage for the other, why did not Volta's mistake—his "repression" of the chemical phenomena of the pile—work in favor of Galvani's camp? Parrot had been one of the first to contest Volta's ideas about the pile, but his challenge had gone unnoticed. In 1829, when the fire of the polemic had been largely consumed and the excitement had abated, Parrot sought an explanation (or consolation) by claiming that after Volta's experiments in Paris "all the physicists embraced the hypothesis with ardent zeal. In Germany, there was something of a propaganda campaign to disseminate it, of which C. H. Pfaff was the self-proclaimed champion. The chemical theory of the pile was eclipsed." (Parrot 1829, 47).

Propaganda is certainly a strong weapon; but, especially in science, it is blunt if not loaded with good arguments. Now, on Volta's side there were at least two arguments, one explicit and highly persuasive, the other usually tacit but decisive. The former appears at the end of Biot's 1801 "Rapport":

That, roughly, sums up Citizen Volta's theory of the electricity known as galvanic. He has sought to reduce all its phenomena to a single one, whose existence is now well ascertained: the development of metallic electricity by the mutual contact of metals. (VO, 2:114)

Carradori used a similar argument:

> The electricity aroused by the motive virtue of dissimilar conductors is therefore sufficient to account for the phenomena in Galvani and Aldini's experiments on animals. One cannot, therefore, resort to another explanatory principle without violating the rules of good philosophy. . . . Galvani and Aldini's experiments do not form a general system, but are particular instances of the general system that Volta has discovered and demonstrated. (1817, 65)

Volta himself had struck the same note by opening his 1814 memoir with the motto, *non plures admittendae sunt causae quam quae verae sunt, et phenomenis expli candis sufficiunt* (one must not admit more causes than those that are true and sufficient to explain the phenomena) (VO, 2:206) and concluding it with the following remark: "Thus, for none of the phenomena of so-called galvanism so far observed, is there any need to resort to an agent other than the electrical—as some have vainly alleged, deliberately creating a new one under the name of galvanic fluid and needlessly multiplying the entities" (VO, 2:297).

Arguments such as these seem to offer a foothold to those who think that scientific disputes can be settled by invoking some pertinent rule of method. For example, one might argue that Volta prevailed because a sound methodological rule prescribes that a theory should be preferred if it solves the problems contained in the competing theory and in others as well. Such a rule, similar to one frequently advocated by Popper (see for instance Popper 1972, 15: let us therefore call it P), can be formulated in the following terms:

$$P: T^* \text{ is preferable to } T \text{ if } Q(T^*) > Q(T),$$

where Q is the quantity of problems (or facts) a theory solves (or explains).

Does P really work in our case? Even though its abstract content may look reasonable, its *application* raises a delicate problem. The rule that one theory is better than another if it explains its own phenomena as well as the other's phenomena—or the rule that one cause is better than two—holds good if we want to explain a single phenomenon rather than two phenomena belonging to two different kinds of domain.[16] But this is precisely the core of the Galvani-Volta controversy. Is the domain to be explained solely physical, solely biological, or physical in the case of the pile and biological in the case of the frog? Raising this problem—whose solution is a prerequisite for the application of P to our case— takes us back to square one. The point to be settled beforehand is to which ontological class the frog phenomena belong—that is, in our terminology, which interpretative theory is more appropriate for them. *P* will not work *unless* this decision is taken; but, owing to the nature of interpretative theories, the decision cannot be taken on the empirical grounds to which *P* refers.

As *P* is not the only sound rule of preference that many serve the methodologist's purpose, let us try a different one. Imre Lakatos's rule of preference (*Lak*, for short) is a sophisticated version of *P* (see Lakatos 1978). It prescribes that, given two theories, we must prefer the one belonging to a research program that is "progressive" with respect to the research program incorporating the other theory. In other words:

> *Lak*: $T^\star$ is preferable to T if $T^\star \subset P^\star$
> and $T \subset P$ and $Q(P^\star) > Q(P)$,

where P⋆ and P are two research programs, and Q the quantity of corroborated novel facts predicted by them.

Applied to our case, *Lak* too favors Volta's theory and seems to fit the historical facts, for most scientists regarded the novel fact of the pile as decisive. *Lak*, however, has the same limits as *P*, for its application raises the same problem. We cannot hold the novel fact of the pile to be decisive unless we are willing to admit that it belongs to the same domain as the frog phenomena. But does it? Once again the question is whether Volta's interpretative theory is better than Galvani's; and once again we cannot obtain the answer using only the empirical means *Lak* takes into consideration.

Larry Laudan's rule (*Lau*, for short) holds greater promise. It states that, when we are faced with two rival theories belonging to two different research traditions, it is rational to prefer the theory belonging to the more progressive tradition, namely, the one with the greater problem-solving effectiveness (Laudan 1977, 68). In other terms,

> *Lau*: $T^\star$ is preferable to T if $T^\star \subset R^\star$ and $T \subset R$
> and $\text{Eff}(R^\star) > \text{Eff}(R)$,

where $R^\star$ and R are two research traditions and Eff is the problem–solving effectiveness, determined by

$$\text{Eff} = E - (A + C_i + C_e),$$

where E, A, C_i, and C_e are, respectively, the quantity of empirical, anomalous, internal, and external conceptual problems a theory confronts. If, as in the Galvani–Volta controversy, $E + A + C_i = E^\star + A^\star + C_i^\star$, *Lau* amounts to prescribing that, in the last resort, the theory to be preferred is the one that best fits the dominant metaphysical assumptions (which are a special source of external conceptual problems). Thus *Lau* differs from both *P* and *Lak* because it is not a mere empirically based rule. This is an important step forward, but it is still not enough. As metaphysical assumptions are typical interpretative theories, *Lau* too cannot become operative until *after* a discussion has taken place and an agreement has been reached as to which interpretative theory is more appropriate. Such a discussion is essential because the mere compatibility of an explanatory theory T^* with the prevailing assumption or interpretative theory I^* does not guarantee that T^* is better than T—which is linked to or compatible with I—unless I^*, after examination, is ruled to be superior to I. This does not mean that *Lau* is useless or unrealistic, only that it offers no operative guidance unless implemented with criteria establishing or weighing the merits of interpretative theories. But what are these criteria, and do they even exist?

If they do, at least some of them must refer to values. Which ones? In a seminal paper, Thomas Kuhn has listed at least five that "provide *the* shared basis for theory choice" (Kuhn 1977, 322): accuracy, consistency, scope, simplicity, and fruitfulness. We could accordingly try to state a new methodological rule in terms of these values. Such a rule (K, for short) would read more or less like this:

K: T^* is preferable to T if $T^* \subset P^*$ and $T \subset P$
and P^* satisfies B better than P,

where P^* and P are two paradigms and B is the basis of the factors involved in theory choice.

What a methodologist might find disappointing in K is that, unlike P, *Lak*, and *Lau*, K is not a genuine, clear-cut rule of method, for it does not clearly specify which particular hierarchy and criteria of values must be applied to the theory choice. But this vagueness in the prescriptive content is precisely K's important feature. K tells us that preference between two rival theories stems from a decision that cannot be regimented by any universal rules. The decision can be taken only in the course of a discussion in a concrete, historical context, aimed at persuading and inducing a consensus in the absence of proofs or decisive, crucial experiments. As Kuhn has remarked, we have to examine "not only the impact of nature and logic, but also the techniques of persuasive argumentation effective within the quite special groups that constitute the community of scientists" (Kuhn 1962, 94).

The "techniques of persuasive argumentation" are those of rhetoric. But rhetoric may induce a consensus when certain conditions are fulfilled—in particular, within science, when an agreement is reached on the set of values forming the shared basis for theory choice, on the hierarchy of these values, on a specific interpretation of the values of such a hierarchy and on a group of commonplaces or presumptions (such as maxims, accepted theories, assumptions taken for granted, and procedural norms) that shift the burden of proof onto the shoulders of the party that violates or modifies them.[17] In the case of Volta's theory, this is the second major argument that worked in its favor, securing its acceptance in the scientific community.

If we take a closer look, the truly decisive value was not the economy of hypotheses (as Volta himself, Biot, Carradori, and many others believed) but the economy of *biological* hypotheses;[18] and the assumption that carried greater weight was not the methodological one of unitary explanation, but the metaphysical one of *physicalist* reduction. We have already seen reductionism at work in Volta's writings, especially in his final explanation of torpedoes and other electric fish. Pietro Configliachi's prefatory note to Volta's last great memoir is an eloquent testimony to the dissemination of this view:

> Who today could be unaware that Physiology and Pathology, now more than ever, are in jeopardy from similar false suppositions; that they are using these as a base to create new erratic systems pertaining to organization, animal functions, and life, with no tangible benefit? On the contrary, such systems work to the detriment of young people, who, through no fault of their own, are still too inexperienced to guard themselves against the lure of novelty and all that is flashy and specious. (VO, 2:210)

A masterpiece of rhetoric, no doubt, implementing an array of seductive devices. Like any concerned father, Configliachi brandishes the carrot and the stick, flattery and threats—a profusion of sympathy toward youth in danger from pseudoscience, a condemnation of serious rival hypotheses as "false suppositions," and a demolition of other research programs branded as "erratic systems." If in 1814 a professor of experimental physics addressed himself to physiologists and biologists with such patronizing arrogance, it was because he felt that reductionist ideology was gaining ground fast. Volta and his followers were touching the right chords: they played on the prevailing physicalist and reductionist beliefs just as skillful lawyers play on a jury's most secret feelings. The technique was bound to pay off.

Some would argue—in the hallowed name of scientific method—that rhetoric in science plays a vicarious, ornamental role at best, and that the Galvani-Volta dispute was ultimately settled by the later experiments of Leopoldo Nobili, Carlo Matteucci, and others. To such objectors, we might reply that the story is somewhat different. Granted, Nobili and Matteucci's experiments did convince the scientific community, but they were not really crucial.[19] Strictly speaking, they too proved the existence of electricity *in* the frog, not of an electricity *of* the frog. One example clearly illustrates this. Having observed that "frogs do not contract under the action of their own current except for a short time, while their current continues to act on the galvanometer for many hours," Nobili concluded that this, "more than any other fact, proves the independence of the electromotive force from those inherent to life" (Nobili 1834, 75). From the very same observation, Matteucci drew the opposite conclusion, namely, that the current's persistence was due to the "organic disposition of the living muscular fiber" (Matteucci 1842, 336). Here too, the difference in the conclusions obviously stems from a difference in the interpretative theories tacitly linked to the premises.

To those who—again in the name of scientific method—object that a preference between rival theories cannot depend on a contingent distribution of epistemic values, we might reply that such a distribution was indeed involved. In their structure, Nobili and Matteucci's experiments were not very different from Galvani's, especially his third experiment. Yet the former convinced the scientific community, while the latter did not. How can we explain this conversion if not by assuming (1) a redistribution of the factors constituting the shared basis for theory choice, and (2) a greater tolerance for the biologistic interpretative theory, favored by the conviction—acquired in the meantime—that Volta's explanatory theory of the pile was untenable?

Lastly, to those who—once more in the name of scientific method—are wary of rhetoric and want to keep scientific rationality in the thrall of methodological rules, we might reply that scientific rationality is too complex a question to be solved by a few methodological formulas, and rhetoric too subtle and serious an art to be reduced to mere propaganda or indoctrination. To be rational is to put forward good arguments; and to put forward good arguments is to appeal to a basis of shared objective factors and an established dialectical framework.[20]

Volta once wrote: "The language of experiment is more authoritative than any reasoning; facts can destroy our ratiocination—not vice versa" (VO, 7:292). The controversy on animal electricity disproved this golden maxim. But it did not disprove that, even beyond the language of facts, one can conduct a rational discussion.

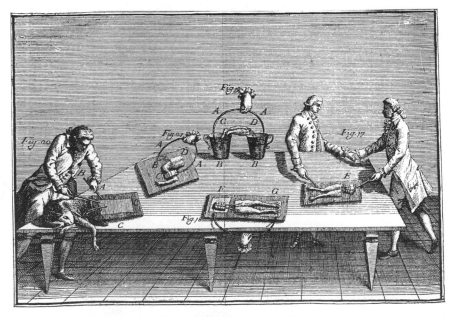

6.6 Plate 4 of Galvani's *Commentarius*.

Notes

Chapter 1
Electricity, the Science of Wonders

1. "'Fräulein' is a word that means nothing else than a marriageable Girl. The term is used in Austria chiefly to denote Girls of the Nobility, and one should not use it incorrectly, as it is regarded as a title of distinction" (Sguario 1746, 40n). These girls, or their mothers, such as the celebrated Donna Laura Bassi, "are not averse to good explanations, but derive endless pleasure from experiments" (Beccaria 1758, 29).

2. In 1749, Giovanfrancesco Pivati lamented that "the extravagant and incredible electrical phenomena have now been shamefully wrested from poor Physics, sold for money by jugglers on public squares, and exposed to the amusement of the ignorant populace" (1749, 13).

3. Hauksbee's machine is described by Newton in the second edition of the *Opticks* (1717) under query 8 (see Newton 1952, 340–41).

4. On this topic see Torlais 1969 and Heilbron 1966a, b.

5. On the diffusion of the Leyden jar, see Heilbron 1979, chap. 13, pts. 1, 2. In Italy, possibly the first description may be that of Sguario (1746, 376), which includes a letter from Leipzig describing the effects of the discharge on the human body.

6. The number, however, is uncertain. Sigaud says 140 (1785, 237); others "about two hundred" (*HAS*, 1746, 11) or even 240 (Barbeu Dubourg 1773, xv); Le Monnier (1746) gives no figure.

7. See Nollet 1746a, 19; 1748a, 164.

8. The quotation is taken from Heilbron (1979, 354), who has examined the manuscript of Nollet's "Journal de voyage de Piedmont et d'Italie en 1749."

9. On the story of the electrotherapy of paralysis, see Hoff 1936; Rothschuh 1960a, 1963; Benguigui 1984, chap. 3, with full bibliography; and Rowbottom and Susskind 1984.

10. Vivenzio (1784, 65) asserts that urban women were more prone to hysteria, "because, by dressing as they do today, they make their bodies liable to attract an excessive quantity of electrical fire in every kind of weather in the atmosphere."

11. Naturally, there was not even a shred of theory to explain why electrical shocks could cure "somber melancholic disorders and hypochondriac ones," still referred to as "the Medici derangement" (Sguario 1746, 386); but when electroshocks were rediscovered and systematically administered nearly two centuries later—on the basis of the theory, if one wants to call it that, of Ladislas von Meduna—the situation was hardly better. For the story of the rediscovery, see Mellina 1981; Rowbottom and Susskind 1984, 193.

12. Franklin (1941, 234) and, in his wake, Beccaria (1753, 132n; 1772, 278) also criticized the use of *intonacature*, which Franklin referred to as the "draw[ing] out of the effluvial virtues of a non-electric . . . and mixing them with the electric fluid."

13. We can therefore justify Marat, who—in the days when he could still climb out of his bathtub on his legs—violently attacked Bertholon's work as being full of "the sheerest absurdities" (see Marat 1784). As for Bertholon, Bachelard (1972, 59) spoke of the "nefarious influence of Baconianism at 150 years' remove."

14. The early experiments on frogs (1792), which showed that they did not contract when charged with a stagnant, noncirculating electricity, led Volta to criticize even "bathtub" electrotherapy (VO, 1:49). Earlier, in a letter to van Marum, Volta had already raised doubts about electrical medicine (VO, 4:71).

15. On the Franklin-Nollet controversy regarding lightning, see Cohen 1956, chap. 11, and Torlais 1956; on the origins of the lightning-rod, see Brunet 1947 and Benguigui 1984, chap. 2.

16. The fullest presentation and critical discussion of Franklin's theory remains Cohen 1956.

17. On this point, Franklin extricated himself from the difficulty with a positivist move à la Newton: "It is not important for us to know the manner in which nature executes its laws; it suffices to know the laws themselves. It is useful to know that saucepans left unsupported in the air fall and break; but to know *why* they fall and *why* they break are theoretical questions. Certainly it is pleasant to know them, but we can save our saucepans even without them" (1941, 219).

18. In this connection, Beccaria too, like Franklin, adopted a positivist stance, explicitly citing Newton's well-known rule in query 31 of the *Opticks* against occult qualities. On Beccaria, see Gliozzi 1935, 1962.

Chapter 2
Volta's and Galvani's Scientific Training

1. Theories of the electrophore were developed in the works of Wilcke and Aepinus; Ingenhousz offered one in 1778. Significantly, Sigaud's *Précis historique* cites Wilcke as the true inventor not only of the instrument but also of the term "perpetual electrophore," and regards Volta as a re-inventor; as for the theory, Sigaud presents that of Ingenhousz (Sigaud 1785, 551–91).

2. In fact query 24, as Beccaria himself had accurately recorded a few pages earlier (p. 126).

3. On pre-galvanic theories of muscular contraction, see Fulton 1926, "Historical Introduction"; Hoff 1936; Cameron Walker 1937; Field 1959, chap. 1; Home 1970; Rothschuh 1960a and 1973, chap. 4.

4. See Robinson 1734, 87–88: "Muscular Motion is performed by the Vibrations of a very Elastick Æther, lodged in the nerves and membranes investing the minute Fibres of the Muscles, excited by the Heat, the Power of the Will, Wounds, the subtle and active Particles of Bodies, and other Causes."

5. See also Laghi 1756 and Marc'Antonio Caldani's second letter to Haller (Haller 1760, 3:461): "Laghi conjectures that the spirits are of an electrical nature." Volta certainly had in mind experiments and considerations such as Laghi's when he wrote, in *Memoria prima* (1792), that the advocates of the theory of the nerveo-electrical fluid "rested it chiefly on the electrical fluid's known consummate ability to irritate muscles. This is shown by the fact that when a muscle of an already dead animal, or of a severed limb, no longer responds to any other stimulus, mechanical or chemical, a small amount of electrical fluid can revive it somewhat and cause it to contract—either by striking the muscle itself with a spark of average intensity, or even without striking it directly, if the fluid is made to flow through the muscle at an adequate speed. From this, they wanted

to conclude that because the electrical fluid was the most efficient of fluids and the sovereign agent of muscular irritation and motion, it was more than likely that nature used precisely this fluid for such a purpose in the animal economy" (VO, 1:22; VOS, 120).

6. Haller's opinion on the nervous fluid was as follows: "If we are asked what we think about the nature of spirits, we shall say that they are an *active* element, highly suited to receiving motion from the will and from sensations; a very fast element, subtler than any fiber of the senses, and yet bigger than fire, aether, electricity and magnetic matter since it can be contained in jars and shackled; lastly, an element that is born of, and is manifestly reinvigorated by, our nutriment" (1766, 381).

7. M. A. Caldani 1757a, 332; see also 1786, 144: "The mere effluvium of electrical current (not the electrical spark) that just barely stirs sentience is the best suited of all stimulants to excite irritability."

8. Priestley's renunciation of inductivist strictures is well documented by the following excerpt, which is worthy of Popper: "In extenuation of my offence, let it, however, be considered, that *theory* and *experiment* necessarily go hand in hand, every process being intended to ascertain some particular *hypothesis*, which, in fact, is only a conjecture concerning the circumstances or the cause of some natural operation; consequently that the boldest and most original experimenters are those, who, giving free scope to their imagination, admit the combination of the most distant ideas; and that though many of these associations of ideas, will be wild and chimerical, yet that others will have the chance of giving rise to the greatest and most capital discoveries; such as very cautious, timid, sober, and slow-thinking people would never have come at" (1775a, 1:258–59).

9. On this subject, Sguario, with his customary lucidity, had previously written that "if some animals acquire electricity when rubbed, it is only through their fur" (1746, 355). Sguario also regarded as inconclusive Etienne Hales's evidence for the electricity of the blood—evidence that Bertholon, for his part, accepted (1780, 144–45). On the misreading of these signs, see Volta (VO, 1:18–21), who in talking about "uncertain or ambiguous experiments" explicitly refers to Bertholon. See also Aldini 1792a, viii, who, after reporting the previous phenomena, stated "minime esse cum inventa animali electricitate confundenda" ("must hardly be confused with the animal electricity discovered").

10. This is also obvious from the actual circumstances in which the book was written. Bertholon submitted it to an essay contest sponsored by the Lyon Academy in 1777 on the topic, "What are the diseases that depend on the greater or lesser quantity of electrical fluid in the human body and what are the means of curing the two kinds?" The work's chiefly practical purpose is also apparent from Bertholon's statement of intent: "Sound principles, founded on experience and observation," he wrote, "form the basis of this book, which discusses electrical hygiene, pathology, and electrical therapy, new sciences whose importance is not open to question" (1780, 4).

11. The *Giornale* is Galvani's diary-record of the protocols of his experiments from 6 November 1780 to 25 April 1787. See GM, 233–411.

12. G. B. Fabri also compiled a three-volume edition of these writings (1757a, 1757b, and 1759).

13. The role of the Bologna environment in Galvani's training and in the origins of galvanism has been examined with great accuracy in Heilbron 1987.

Chapter 3
Galvani's Experiments and Theory

1. The experiment of contractions at a distance is referred to as the "first experiment" in Fulton 1940, 302. But in Fulton 1926, 36, the same author speaks of "Galvani's (first) experiment" to denote the experiment with metals—which I shall refer to as the "second experiment." My numbering follows Hoff 1936 and Dibner 1952, who, however, do not examine the "fourth experiment" (see chapter 6, §6.1).

2. See Stanhope 1779, 113–14: "Instead of an electrified Prime-Conductor, let us suppose a *real Cloud* ABC charged with *Electricity*. And let us suppose, that there be *Persons*, standing either insulated or *not* insulated, either near each other, or near some conducting Body or Bodies of any kind whatever; provided those Bodies be *not Conductors* well connected with the Earth, and terminating at their upper Extremity in *prominent†* Points: [author's note: †The reason for *this exception* will be seen hereafter] then, I say, that those *Persons*, if they be *superinduced* by the Thunder-Cloud's *electrical Atmosphere*, may (under circumstances similar to those above-explained,) receive that kind of *electrical shock*, which I have, in the foregoing Pages, called a *returning Stroke*: and that, if those Persons be *strongly superinduced* by the *electrical Atmosphere* of the Cloud, they may (under circumstances similar to those explained above,) receive a very *strong shock*, be knocked down, or be even *killed*, at the Instant that the Cloud discharges, with an *Explosion*, its Electricity; whether the Lightning falls *near the very Place* where those Persons are, or at a *very considerable Distance* from that Place; or whether the Cloud be *positively* or *negatively* electrified.

"So that, a *Person*, placed (for instance) upon the Surface of the Earth at F, and strongly *superinduced* by the *electrical Atmosphere* of the Cloud *ABC*, might receive a violent *returning Stroke*; even, if the *main Explosion* (which produces the *returning* Stroke) were to take place, at the most *remote* Extremity C of that Thunder-Cloud."

3. Galvani listed other reasons in the days that followed, for example, on 3 February in "Exp. no. 2" and the "Introductory Note" to "Exp. no. 4" (GM, 260–61).

4. On this influence, see Home 1970.

5. On this issue, see Gherardi 1841a, 74–81. It seems, however, that in his justified attempt to refute the notion of Galvani's ignorance—indeed, his happy ignorance—Gherardi overstated his case by claiming, for example, that Galvani was undoubtedly aware of the returning-stroke theory. In this, Gherardi failed to distinguish the period of the first experiments on contractions at a distance from the later period of the *Commentarius* and *Trattato dell'arco conduttore*.

6. The anonymous author S of the *Transunto* (partial Italian translation of the *Commentarius*), probably Lazzaro Spallanzani, commented: "This explanation is in fact the most consistent with the modern theories; and if no contractions are produced with the discharge of the magic square, it is because the electricity, in this case, merely moves from one surface of the square to the other, without inducing the said change in the aerial strata" (S 1792, 21n).

7. On interpretative theories, their status and function, see Pera 1982, 1988b.

8. This awareness is clear in the *Trattato dell'arco conduttore* of 1794. Referring to the designation of the innervated muscle as a "miniature animal Leyden jar," Galvani wrote in the third person: "But if for other reasons he wished to refer to the muscle thus— chiefly because of the equivalent effects he observed between the excitation of muscular

contractions and the release of the brush discharge on a Leyden jar when the spark is drawn from the electrical machine or another charged jar—I could not concur with him so readily, nor approve his hypothesis and designations; for the equivalence of effects might perhaps be just as well explained by what modern physicists call the returning stroke, whose mechanical impact stimulates the nerve and thus activates animal electricity" (GO, 206; GOS, 391–92). It will be noted, however, that even in this passage the returning stroke was construed as acting via a "mechanical impact": in other words, the atmospheres remained material. Regarding the appearance of the brush discharge on the Leyden jar when a conductor was discharged nearby, see below, especially note 20.

9. On the date, see Gherardi 1841a, 100–101.

10. Alibert commented: "No doubt it is astonishing that so many preliminary attempts were required before reaching this point. So it is that the human mind is condemned to languish in science's well-trod paths before attaining the most valuable discoveries" (1802a, 59).

11. The *Giornale* also refers to an iron hook (GM, 392). In the *Commentarius*, however, the hook is said to be of brass (GM, 118; COS, 262; GF, 59) As we shall see, the detail is not irrelevant

12. So says Galvani in the *Commentarius*, but in his already-quoted paper of 30 October 1786, he writes: "We very nearly thought we had found what we were looking for" (GM, 33; GOS, 163).

13. This is "Galvani's galvanism," which Mamiani (1983, 27) regards as the "focal motive" of the entire subsequent controversy with Volta.

14. See the *Commentarius*: "These results . . . led us to suspect . . ." (GM, 119; GOS, 262; GF, 60).

15. According to the *Transunto*, this was "the best and most entertaining proof" (S 1792, 8).

16. For this experiment, see Galvani's precautions as described by Aldini in his notes to the *Commentarius* (1792b, 272n).

17. This hypothesis was also confirmed by experiments with dissimilar fluid armatures, such as those described by Aldini (1792b, 266n), who repeated a similar test by Galvani (GM, 141; GOS, 280). Even fig. 6.6, exhibit 19 (Table 4 of the *Commentarius*) shows a striking resemblance between this experiment and Volta's "crown of cups" pile. See the convincing reconstruction by Tabarroni 1971b.

18. The term is used in Galvani's 1786 paper already quoted, and in the one entitled *Electricitas animalis* of 1787 (see GM, 61; GOS, 196).

19. Alibert's summary is presumably the source for the one provided by du Bois-Reymond 1848, 1:49, English trans. in Hoff 1936, 159, cited by Dibner 1952, 17–18, and Cohen 1953, 28. In his discussion of Galvani's views, Hoff later notes: "It is readily seen that in proposing the brain as the source of animal electricity by *separation* from the blood, Galvani simply adopted the current theory that animal spirits were distilled off from the blood in the brain, and substituted animal electricity for animal spirits" (1936, 167; italics mine). Interestingly, the contemporary Italian translation of Alibert (1802b) renders "sécrété" (secreted) in point 2 as "separato" (separated).

20. This last point, in turn, was the source of other analogies. It was based on the phenomenon—which Galvani may have been the first to discover in 1786—of the brush discharge that reappears on the Leyden jar when a spark is drawn from a charged conductor placed nearby (GM, 156; GOS, 293; GF, 75). The most significant analogy de-

rived was the following: in the case of the brush discharge, the discharge of the conductor seems to excite the jar's internal electricity; by analogy, in the case of the frog, we may suppose that the muscle is induced to discharge its own intrinsic electricity—thus producing contractions—when a conductor is discharged nearby. In the *Commentarius*, Galvani, calling this brush-discharge phenomenon "a significant new argument for the analogy we have already put forward," claims to have noticed it "through a chance observation" and "recently" (GM, 156–57; GOS, 293; GF, 75). It is more likely, however, that he observed it at the time of the experiments described in the paper, *De consensu et differentiis inter respirationem et flammam, penicillumque elettricum prodiens ex acuminato conductore leidensis phialae de industria oneratae*, which, as Tabarroni has shown (1971a, 151n), dates from 1786. In that case, the analogy between the brush discharge and muscle might be the original inference behind Galvani's discovery of animal electricity. On the role of analogy and on the logical reconstruction of that discovery, see Pera 1987a.

Chapter 4
Volta's First Reaction

1. Carminati immediately informed Galvani of Volta's opinion; in his reply Galvani restated his theory, albeit dubitatively (GO, 135–40; GOS, 321–24).

2. See *Memoria seconda*, §§13–14 (VO, 1:46–47; VOS, 392–93); also VO, 1:108–9, 115.

3. Like the first, it appeared in Luigi Brugnatelli's *Giornale fisico-medico* in Pavia, but in two parts; the second part (secs. 50–96) was published in July 1792.

4. See also experiments A and B referred to by Volta in the first letter to Tiberio Cavallo of 13 September 1792 (VO, 1:181–82).

5. See the discussion of H_6 in §3.3 above. For Galvani's experiments on contractions with the arc applied to the nerve alone, see the discussions by Gherardi (1841b, 465–68) and Aldini (1792a, 228).

6. Sulzer's experiment, quite casually referred to in a note to the third part of his prolix *Nouvelle théorie des plaisirs*, was the following: "If we join two pieces, one of lead, the other of silver, so that the two edges form a single surface, and if we bring them up to the tongue, we will feel a taste sensation fairly close to that of iron sulfate, whereas each piece taken separately will produce no trace whatever of such sensation. It is unlikely that this conjunction of metals causes an outflow of a solution from either one, and that the dissolved particles penetrate the tongue. We must therefore conclude that the conjunction of these metals causes in either one, or in both, a vibration of their particles, and that this vibration, which must necessarily strike the nerves of the tongue, produces the sensation of taste there" (1767, 155–56n).

Volta did not know of Sulzer's experiment (VO, 1:152; VOS, 425), and, in any event, it is hard to fault him when he argued that Sulzer's explanation was "hypothetical, vague, and unfounded" (VO, 2:279n).

7. See also VO, 1:265: *Nuova memoria sull'elettricità animale*, first letter, 10 February 1794.

8. This circularity flaw is even clearer in the following passage, in which Carradori, recounting the controversy many years later, sought to support Volta's point of view: "Any accidental factor that induces a change in the substance can produce the *heterogeneity* of the metal—such as the accidental properties of *hardness, temper, smoothness, shine,*

heat, etc.; and when a metal of the same species produces the effects of dissimilar metals, one must say that an accidental quality has caused it to suffer changes that have made it *heterogeneous* or *dissimilar*" (1817, 23–24).

9. Aldini 1792a, 236.

Chapter 5
The Crucial Experiments

1. For details of these opposing camps, see Alibert 1802a, 76–79; Sue 1802; du Bois-Reymond 1848, vol. 1, pt. 1, chap. 1; Figuier 1868, 612–19; Sirol 1939, 49–50.

2. Spallanzani also pronounced a very harsh and ungenerous judgment on his colleague Volta. In a letter to the Abate Paolo Spadoni of Bologna of 29 March 1794, he wrote: "Today, Alessandro Volta, awarding degrees to some Engineers, read a long, long speech, entirely directed against Signor Galvani's electricity. In it, he claimed to prove that this electricity should not be properly be called *animal*, but rather *metallic*, as he considers it to be exclusively generated by the armatures. He based himself on several of his experiments, which, as is his wont, were awash in a sea of words. But for us—his colleagues—who listened to him, he failed to dispel our factually-based opinions in favor of a truly animal electricity. I believe he will shortly publish his address. It will resemble his other papers. This worthy Colleague of mine has an inventive spirit, but his head is full of shaky hypotheses, he lacks an observer's logic, and he is always [wordy?] in expounding his ideas" (1964, 32).

3. Author of eleven letters on animal electricity and *Experiments on Animal Electricity* (1793a). On Valli's role, see Coturri 1968.

4. See Vassalli-Eandi 1803a, 1803b; and VO, 1:290; VOS, 455.

5. See Humboldt 1799, chap. 10. According to Volta (VO, 2:218n; VOS, 606n), Humboldt eventually changed his mind. It is more likely, however, that he maintained his independent position. See Rothschuh 1960b.

6. Fowler argued—against Galvani and especially against Valli—for the need for dissimilar metallic contacts. For example, he wrote: "I found that I could not excite in an animal the appearances described by Galvani with any substances whatever, whether solid or fluid, except the metals; and that the mutual contact of two different metals with each other, so far as I was able to determine, was in every case necessary to the effect" (1793, 4). Fowler also wrote that the electrical fluid—or "influence," as he called it—discovered by Galvani "has no relation whatever to electricity" (p. 164). Repeating Sulzer and Volta's experiment, he did find "a considerable difference" between the "sensation produced upon the end of the tongue, by coating its upper and under surfaces with different metals," and the sensation "produced by electricity" (p. 82).

At the end of his work, Fowler added a letter addressed to him by John Robison, professor of natural philosophy at Edinburgh, who also reported some experiments on the electrical taste. One of these—performed with "a number of pieces of zinc made of the size of a shilling . . . made . . . up into a rouleau, with as many shillings" and applied to the tongue—is noteworthy because it anticipates Volta's pairs (Robison 1793, 173). Dibner 1952, 35, has rightly drawn attention to this experiment.

7. For Volta's first ranking, see VO, 1:65–66n; VOS, 417–18n. A later ranking is in VO, 1:304; VOS, 454.

8. In the last edition of his famous *Treatise*, Cavallo reported Galvani's and Volta's

experiments but did not take a position with respect to the controversy (Cavallo 1795, 3:1–75).

9. Anonymous, like the subsequent *Supplemento*. Receiving the latter, Spallanzani wrote to Galvani (10 November 1794) that he believed it "to be your work, as I have also firmly assumed the *Arco conduttore* treatise to be"; and he added: "If I am mistaken in my opinion, I shall at least have the pleasure of seeing my mistake shared by all those who will read these two small Works. Both are too masterly to be attributed to another author. These further attempts thus add a new luster to the previous ones, and give greater prominence and extension to your sensible theories" (1964, 70). For the authorship of the *Trattato*, the decisive evidence is Galvani's letter of November 1793 to Mariano Fontana, in which the book is announced before publication as the work of "an unknown author" (Galvani 1938, 75). See also Spallanzani's letter of 10 September 1794 (Galvani 1938, 84–85) and the discussion by P. Di Pietro (ibid., 86–88).

10. This is a reference to Volta, who in the second letter of *Nuova memoria sull'elettricità animale* had written that "all the magic . . . resides in the bodies of the metallic class" (VO, 1:279).

11. In *Memoria prima*, Volta identified "four degrees or stages of death" (VO, 1:31; VOS, 381). Galvani, instead, distinguished between three states of force in the animal "since it proves too tricky and difficult to establish any more, as a learned Author has done" (GO, 159; GOS, 368). The allusion is to Volta.

12. This is Aldini's experiment, reported here on pp. 119–121.

13. It is here that Galvani backtracked, conceding to Volta that the muscle–jar analogy did not apply to the phenomenon of contractions at a distance (his "first experiment"), which Galvani too now explained with the returning stroke (see §3.1 above).

14. Curiously, while this opinion collapsed, another of Volta's notions was revived. In the last chapters of the *Trattato*, Galvani introduced some changes in the physiological theory of muscular contractions that he had set out in the last part of the *Commentarius*. In particular, he rejected the idea that the contractions were due to a mere flow of fluid; now, he accepted the hypothesis that they occurred when "one alters the system of this current, to wit, when one alters its peaceful, even course through the natural arc and increases its velocity"—precisely Volta's opinion in *Memoria prima*.

15. Or else it succeeded "by a miracle" (VO, 1:316–17).

16. Humboldt thought so. See 1799, 45, quoted p. 121 above, and the further comment: "Volta . . . replies to the experiments performed with mercury that there is a great difference between the surface of this metal and the inside of its mass, because the surface oxidizes in contact with the atmosphere; so that, in Aldini's experiment, the conducting arc is homogeneous only in appearance, since the frog's organs are immersed at different depths; and the mercury, in these experiments, produces different impacts at each end of the arc, resulting in an uneven development of electricity. Thus, against the phenomena described by Aldini, the physicist of Pavia can put up only hypothetical rebuttals" (p. 52).

Spallanzani, as usual, was even stricter against Volta. He wrote to Galvani: "Going back to metals now, and to their dissimilarity—used as a counter-argument—you destroy this imagined pretext with the homogeneity of mercury. When this very felicitous attempt by Professor Aldini was published and became known in Pavia, I could not help laughing when I heard your adversary claim that this metal was heterogeneous—I who was then engaged, as since, in experiments requiring the pneumatic mercury machine, and who saw how slowly that extremely superficial oxide formed on it. But the commitment to support his theory dragged him, in a certain way, into taking liberties with the truth, or rather with the evidence" (1977, 170).

17. See Valli 1794, iv (experiment I) and v–vi (experiment III); GO, 212.

18. See Cima's opinion (1846, 17–18): "In order to support his theory, which was so hindered by Galvani's incontrovertible experiments, Volta certainly overemphasized the weakness of the contractions obtained without the mediation of conducting arcs, and the difficulty of producing such contractions. But the experiment speaks too clearly. Galvani obtained the contractions without wetting the frog; he used a sea-salt solution in only a very few cases, precisely to avert any potential objections. Even if the prepared frog is washed and dried before undergoing the experiment, it will be seen to contract—provided it is resilient and lively. This is a fact anyone can repeat and verify. The contractions obtained by means of the sole contact between the nerve and muscle of the galvanic frog, with no other contrivance, is a constant phenomenon—as constant, of course, as a fact of such complexity can be, and provided the frog meets the requirements; and in force and violence, the contractions match those obtained with dissimilar metallic arcs. It is one thing to say, as Galvani himself and Volta too later recognized, that when the animal's forces weaken, and the contractions accordingly slacken, these are revived by wetting the parts to be joined with various animal liquids, or acid, alkaline, or saline liquids. It is another thing to say that Galvani's frog does not contract without these artifices." Yet it must be noted that Galvani himself, in the *Commentarius*, wrote that "a muscle fiber . . . is composed of solid as well as *fluid* parts (substances which produce in it no slight diversity)" (GM, 154; GOS, 292; GF, 74; italics mine).

19. Volta claimed to have "entertained these notions from the outset" (VO, 1:297; VOS, 465). We have seen that this is true, but Gliozzi 1937, 2:121, rightly commented: "And so? If he had believed in the electrical action—albeit weak—of the contact between second-class conductors, why should Volta ever have had to deny any importance to the experiments of 'Anonymous' (Galvani)? Why resort to the mechanical explanation?"

20. See *Nuova memoria*, first letter (VO, 1:264–65): "But how can we make sure the metals we use are perfectly and completely identical? They may be so in name and substance, but may differ considerably in their accidental qualities such as hardness, temper, surface polish and shine, heat, *etc.*" In the second letter of the same memoir (VO, 1:275n; VOS, 442n): "An accidental difference in temper, polish, *etc.* is enough to produce to some degree the effects mentioned." Letter to Orazio Delfico (13 April 1795): ". . . dissimilar, either in substance—such as gold, silver, and mercury on the one hand, and lead, tin, and zinc on the other—or even solely because of some accidental diversity of temper, polish, *etc.* Or a dry conductor between two equally different humid ones, such as pure water, salt water, vinegar, spirit of wine, ink, liquors, acids, alkalis, *etc.*, milk, serum, mucus, saliva, urine, bile, etc." (VO, 1:338). Volta added that the conductors had to be "*in some way* dissimilar" (ibid.; italics mine in all quotations).

21. This capacity to adapt post factum, which was precisely a weakness of Volta's theory, was presented instead by Arago as a strength. In a historical and logical exaggeration, he wrote: "Volta had dealt a death blow to animal electricity. His conceptions constantly adapted themselves to the poorly interpreted experiments that—so it was hoped—would help to undermine them" (1854, 217).

22. Volta's situation was well described by Carradori (1817, 31), who observed that Volta had not yet "managed to bring [his electricity] into general view, as common

electricity or universal electricism—in whatever manner reawakened—does through a few artifices."

23. See VO, 1:381 (second letter to Mocchetti), stating the difficulties; and VO, 1:405 (first letter to Gren), in which Volta supplies a ranking.

Chapter 6
The Pile

1. It is usually regarded as the fundamental experiment of electrophysiology; see Cima 1846, 28; du Bois–Reymond 1848, 1:84; Pupilli 1956, 449; Pupilli and Fadiga 1963, 575; Moruzzi 1964, 114.

2. The reference is to the passage in Galvani's text quoted p. 146 above: "He, in short, attributes everything to metals."

3. It should also be noted that Volta's compromise hypothesis, in support of which Volta himself cited the case of the torpedo, was untenable, in Galvani's view, because of the lack of contractions when the torpedoes were decapitated.

4. References are given to the French original, reproduced in VO, 1:563–87, and to Volta 1952 (facsimile of the *Philosophical Magazine* translation, probably by William Nicholson, in Dibner 1952, 42–49).

5. Another omission is even more suspect. Volta was not only a prolific writer, but an autobiographical one. Besides faithfully reproducing his protocols, his scientific papers almost always give a thorough account of the genesis of his ideas. But the discovery of the pile is a striking exception. Volta does not even say if he began by building the column pile (as is fairly likely) or the crown-of-cups pile. One cannot help but suspect that his discretion was intended to avoid too close a comparison between the crown of cups and the similar experiment by Galvani and Aldini, reported and illustrated in the *Commentarius*. See fig. 6.6, exhibit 19, and n. 16 to chap. 3, which lists the study by Tabarroni that points out the resemblance.

6. Heilbron—who has painstakingly examined the experiments that led Volta to construct the pile—very aptly remarks: "The pile was the price of [Volta's] stubborn refusal to see anything but the operation of common electricity in effects ascribed to other agencies by the Galvanists." See Heilbron 1977, 60. Gill sees in the attack on the Galvanists "the reasons for the long interval between the experiments in which the elements of the two types of battery were discovered, and their combination to form the useful versions of the *Couronne de tasses* and the *Pile*" (1976, 368). We must not forget, however, just how central the dispute was for Volta: without it, the pile would never have been invented. Indeed, Volta regarded his discovery as the living refutation of his opponents' theory.

7. Immediately after the publication of the *Commentarius*, Fabroni presented at the Florence Academy in 1792 a chemical theory to explain the contractions of Galvani's frog—but this had no impact on the two opposing camps. In 1799 Fabroni returned to the issue, but with no greater success. Aldini too referred to a "possible chemical nature of the phenomenon" (1792a, xi; GOS, 224).

8. In his 1987 essay—which I was able to take into account for this edition of my book—Kuhn explicitly discusses Volta's case. He distinguishes between two arrangements of the pile elements (pp. 12–13): (a) Volta's original arrangement, zinc/silver/wet blotting paper; (b) the modern arrangement, silver/wet blotting paper/zinc. Kuhn argues

that the shift from (a) to (b) marks the shift from the theory of contact to the chemical theory—a revolutionary change, since it exhibits three crucial features: it is holistic, it alters the set of objects to which the descriptive terms attach, and it modifies the old pattern of similarities.

Kuhn is right, but I find these three features even more evident in the shift from Galvani's theory to Volta's general theory. As I try to make clear in the text, the arrangement of elements does *not* change in this shift; only their reading (the interpretative theory) does, altering the ontological reference class. For this reason the two conceptual systems—Galvani's explanatory theory and Volta's explanatory theory, with their respective interpretative theories—are observationally equivalent. I do not know if Kuhn accepts this point of view and would be willing to recognize that the true revolutionary change—in Kuhn's original sense of a full-scale gestalt switch—was not the shift from the theory of contact to the chemical theory, but the shift from the theory of intrinsic, natural animal electricity to that of physical, artificial contact electricity.

9. See Kipnis 1987, 117: "For the physiologist Galvani a frog was the objective of his study, while the physicist Volta saw in it only a sensitive instrument. Galvani was concerned with a discovery of what makes life *different* from inanimate nature, whereas Volta was interested in finding their common features."

10. Recenti rightly speaks of different "lines of research" that "played a continuous role in guiding the inferences, analogies, and explanations of the situations" (1983, 8).

11. This was, however, Carlo Matteucci's opinion: "Volta then triumphed over his opponent with a famous experiment in which he proved that two metals in such conditions produced electricity, now visible not only in the frog, but also on the ordinary gold-leaf electroscope" (1867, 14). Even Wolf (1961, 1:260) regards the pile as a crucial proof. Kipnis, instead, specifically denies this: "A nerve-muscle preparation can be harmlessly excluded from a circuit with a pile, but not from the circuit without a pile, where it is the only indicator of electricity. But maintaining animal parts in a circuit leaves the possibility of interference by animal electricity. Thus, experiments with a pile cannot at all advance the solution of Galvani's original problem" (1987, 136). I am not sure, however, that Kipnis regards Galvani and Volta's theories as observationally equivalent, for although he writes that "it was hardly possible to decide between the two theories on the basis of the number of phenomena being explained," he also maintains that "Volta's third theory [our V_g, p. 112] would have given him the advantage, had he proved it in all its generality" (p. 124).

12. As for Spallanzani, see what he wrote to Galvani (1977, 169–70): "And if the electricity is made more spirited by means of metals, you excellently argue (and by sound logic one could not argue otherwise) that the metals do not create this electricity, but only augment its intensity."

13. Part of this closing section is based on Pera 1989.

14. Figuier 1868 remains a valuable source. See now Kipnis 1987.

15. These views were notably expounded in Vassalli 1803b and Aldini 1803 and 1804.

16. Kipnis (1987, 130) rightly comments that "since physics did not provide enough evidence to resolve the dispute on the nature of galvanism, scientists turned to metaphysical arguments. An explicit and very popular one was the call for minimizing the number of causes." He also maintains that "two-causes theory of galvanic phenomena would have been too complicated for developing at that time" (p. 124). My view is that the complication is due not only to technical reasons but to philosophical ones. Developing

a two-causes explanatory theory presupposes a tolerance toward the two contending interpretative theories, whose rivalry could not then—and still cannot today—be settled on a technical basis.

17. Some of the substantive and procedural factors of scientific dialectics are examined in Pera 1987b and 1991.

18. This issue of the controversy has now been clearly identified by Kipnis: "It seems that it was Volta's reductionism rather than physical arguments that precluded him from admitting the possibility of an electricity which could exist only in animal bodies" (1987, 131).

19. On the role of Nobili and Matteucci's experiments in vindicating Galvani's view, see Pera 1988b, where I also discuss other epistemological aspects of the Galvani-Volta controversy.

20. In Pera 1988a, I question the link between methodology and scientific rationality, suggesting rhetoric as a substitute for the former.

References

Aepinus, Ulrich Theodor

1756   "Quelques nouvelles expériences électriques remarquables." *Mémoires de l'Académie des Sciences*, Berlin, 12(1758):105–21.

1759   *Tentamen theoriae electricitatis et magnetismi*. Petropoli (Saint Petersburg): Typis Academiae Scientiarum.

1979   *An Essay on the Theory of Electricity and Magnetism*. English trans. by P. J. Connor, intr. by R. W. Home. Princeton, N.J.: Princeton University Press.

Aldini, Giovanni

1792a   *De animalis electricae theoriae ortu atque incrementis*. In Galvani 1792, iii–xxvi; partial Italian trans. by E. Benassi in GOS, 222–37.

1792b   "Note" to Galvani 1792. Italian trans. by E. Benassi in GM and GOS.

1794   *De animali electricitate dissertationes duae*. Bononiae (Bologna): Ex Typographia Instituti Scientiarum.

1802   *Saggio di esperienze sul galvanismo*. Bologna: A S. Tommaso D'Aquino.

1803   *An Account of the Late Improvements in Galvanism*. London: For Cuthell and Martin.

1804   *Essai théorique et expérimental sur le galvanisme*. 1-vol. and 2-vol. eds. Paris: Fournier et Fils.

Alibert, Jean-Louis

1802a   *Eloge historique de Louis Galvani*. Paris: Richard, Caille et Ravier, an X (Year X, 1801–2).

1802b   *Elogio storico di Luigi Galvani*. Italian trans. of 1802a. Bologna: A S. Tommaso D'Aquino.

Ampère, André-Marie

1958   *Théorie mathématique des phénomènes électro-dynamiques uniquement déduite de l'expérience* (1827). Pref. by Edmond Bauer. Paris: Librairie Scientifique Albert Blanchard.

Arago, Jean-François-Dominique

1854   "Alexandre Volta." In *Oeuvres complètes*, 1:187–240. Paris: Gide et Baudry.

Bachelard, Gaston

1972   *La formation de l'esprit scientifique* (1938). Reprint. Paris: Vrin.

Barbeu Dubourg, Jacques

1773   "Parallèle des Théories de Franklin et de Nollet." In Franklin 1773, 1:335–37.

Beccaria, Giambattista

1753   *Dell'elettricismo artificiale e naturale libri due*. Turin: Nella Stampa di Filippo Antonio Campana.

1758   *Dell'elettricismo. Lettere dirette al Chiarissimo Sig. Giacomo Bartolomeo Beccari*. Colle Ameno in Bologna: All'Insegna dell'Iride.

1767   "De electricitate vindice . . . ad Beniaminium Franklinium . . . Epistula." In Franklin 1970, 42–49; English trans. pp. 49–57.

1769a   *Experimenta atque observationes, quibus electricitas vindex late constituitur atque explicatur*. Augustae Taurinorum (Turin): Ex Typographia Regia.

Beccaria, Giambattista (*cont.*)
1769b   *De atmosphaera electrica.* Turin; *Philosophical Transactions* 60(1770):277–301.
1772   *Elettricismo artificiale.* Turin: Nella Stamperia Reale.
1793   *Dell'elettricismo. Opere del G. Beccaria delle Scuole Pie, con molte note nuovamente illustrate.* 2 vols. Macerata: Dalla Nuova Stamperia di Antonio Cortesi.

Belloni, Luigi, ed.
1963   *Essays on the History of Italian Neurology.* Milan: Istituto di Storia della Medicina.

Benassi, Enrico
1963   "Ipotesi elettropatogeniche e proposte elettroterapiche nell'opera di Luigi Galvani." In Belloni 1963, 131–38.

Benguigui, Isaac
1984   *Théories électriques du XVIIIe siècle. Correspondance Nollet-Jallabert.* Geneva: Georg Editeur.

Bertholon, Nicolas
1780   *De l'électricité du corps humain dans l'état de santé et de maladie.* Paris: Didot.

Bertrand, Bernard-Nicolas
1756   *Elémens de physiologie.* Paris: Guillaume Cavelier.

Bianchini, Giovanni Fortunato
1749   *Saggio d'esperienze intorno la medicina elettrica.* Venice: Presso Giambatista Pasquali.

Boerhaave, Hermannus
1743   *Praelectiones academicae.* Editio prima veneta, vol. 3. Venetiis (Venice): Apud Simonem Occhi sub signo Italiae.

Bose, Georg Mathias
1745   "Abstract of a Letter from Monsieur De Bozes." *Philosophical Transactions* 43:419–21.

Brazer, Mary A.B.
1963   "Felice Fontana." In Belloni 1963, 107–16.

Brunet, Pierre
1947   "Les origines du paratonnerre (Discussions et réalisations)." *Revue d'histoire des sciences* 1:213–53.

Caldani, Floriano
1792   *Riflessioni sopra alcuni punti di un nuovo sistema de' vasi assorbenti ed esperienze sulla elettricità animale.* Padua. The "esperienze" (experiments) are on pp. 113–83.
1794   *Osservazioni sulla membrana del timpano e nuove ricerche sulla elettricità animale.* Padua. The "nuove ricerche" (new researches) are on pp. 48–197.
1795   *Lettera nella quale si esaminano alcune riflessioni circa le nuove ricerche sulla elettricità animale.* Padua: Della Stamperia Penada.

Caldani, Marc'Antonio
1757a   "Lettera sull'insensitività ed irritabilità di alcune parti degli animali scritta al chiarissimo, e celebratissimo Signor Alberto Haller." In Fabri 1757a, 269–336.
1757b   "Sur l'insensibilité et l'irritabilité de Mr. Haller. Seconde Lettre [30 December 1757]." In Haller 1760, 3:343–485.
1759   *Lettera terza sopra l'irritabilità e insensitività halleriane al chiarissimo Signor Dottor Gaetano Rossi* (Bologna, 20 June 1759).
1786   *Institutiones physiologicae.* Editio tertia italica, aucta & emendata. Venetiis (Venice): Sumptibus Jo. Antonii Pezzana.

Caldani, Marc'Antonio, and Spallanzani, Lazzaro
  1982   *Carteggio (1768–1798)*. Ed. G. Ongaro. Trento: Istituto Cisalpino Editore, La Goliardica.
Cameron Walker, W.
  1937   "Animal Electricity before Galvani." *Annals of Science* 2:84–113.
Canton, John
  1753   "Electrical Experiments, with an Attempt to Account for their several Phaenomena." In Franklin 1941, 293–301.
Carradori, Giovacchino
  1793   *Lettere sopra l'elettricità animale al Sig. Cav. Felice Fontana.* Florence: Nella Stamperia di Luigi Carlieri sulla Piazza de' Pitti.
  1817   *Istoria del Galvanismo in Italia.* Florence: All'Insegna dell'Ancora.
Cavallo, Tiberio
  1779   *Trattato completo d'elettricità teorica e pratica con addizioni e cangiamenti fatti dall'autore.* Florence: Per Gaetano Gambiagi. This Italian trans. of *A Complete Treatise* contains material not included in the English editions.
  1781   *An Essay on the Theory and Practice of Medical Electricity* (1780). 2d ed. London: Printed for the Author and sold by P. Elmsly, C. Dilly and J. Bowen.
  1782   *A Complete Treatise on Electricity, in Theory and Practice; with Original Experiments* (1777). 2d ed. London: C. Dilly and J. Bowen.
  1784   *Teoria e pratica dell'elettricità medica.* Italian trans. of Cavallo 1781. Naples: Nella Stamperia Regale. Contains Vivenzio 1784.
  1795   *A Complete Treatise on Electricity.* 4th ed., 3 vols. London: C. Dilly.
Cavendish, Henry
  1776   "An Account of Some Attempts to Imitate the Effects of the Torpedo by Electricity." *Philosophical Transactions* 66:196–225.
Cigna, Giovan Francesco
  1765   "De novis quibusdam experimentis electricis." *Miscellanea Turinensia* 3(1762–65):31–72.
Cima, Antonio
  1846   *Saggio storico-critico e sperimentale sulle contrazioni galvaniche.* Cagliari: Tipografia di A. Timon.
Cohen, I. Bernard
  1941a "Introduction" to Franklin 1941, 1–161.
  1941b "The Lectures and Dicoveries of Franklin's collaborator Ebenezer Kinnersley." App. 1 to Franklin 1941.
  1953   "Introduction" to GF.
  1956   *Franklin and Newton.* Memoirs of the American Philosophical Society; new ed. Cambridge, Mass.: Harvard University Press, 1966.
Coturri, Luigi
  1968   "La posizione di Eusebio Valli nella disputa intorno alla così detta elettricità animale sorta tra il Galvani e il Volta, e l'importanza dei suoi interventi per i successivi esperimenti del Volta stesso." *Pagine di Storia della Medicina* 12(5):9–15.
Deluc, Jean-André
  1804   *Traité élémentaire sur le fluide électrogalvanique.* 2 vols. Paris: Chez la Veuve Nyon.

Dibner, Bern

    1952   *Galvani-Volta: A Controversy that Led to the Discovery of Useful Electricity.* Norwalk, Conn.: Burndy Library.

du Bois-Reymond, Emil

    1848   *Untersuchungen über thierische Elektricität.* Berlin: Reiner, 1848 (vol. 1); 1849 (vol. 2, pt. 1); 1860 (vol. 2, pt. 2, folios 1–24); 1884 (vol. 2, pt. 2, folios 25–37).

Du Fay, Charles

    1733a   "Des Corps qui sont le plus vivement attirés par les matières éléctriques, & de ceux qui sont les plus propres à transmettre l'Electricité." *MAS*, 233–54.

    1733b   "De l'attraction et la répulsion des corps éléctriques." *MAS*, 457–76.

    1734   "A Letter from Mons. Du Fay concerning Electricity." *Philosophical Transactions* 38:258–66.

Fabri, Giacinto Bartolomeo, ed.

    1757a   *Sulla insensitività ed irritabilità halleriana. Opuscoli di varj autori raccolti da G. B. Fabri, Parte prima.* Bologna: Per Girolamo Corciolani ed Eredi Colli e S. Tommaso D'Aquino.

    1757b   *Sulla insensitività ed irritabilità halleriana. Opuscoli di varj autori raccolti da G. B. Fabri, Parte seconda.* Bologna: Per Girolamo Corciolani ed Eredi Colli e S. Tommaso D'Aquino.

    1759   *Sulla insensitività ed irritabilità halleriana. Supplimento agli Opuscoli di varj autori raccolti et in due parti diviso da G. B. Fabri.* Bologna: Per Girolamo Corciolani ed Eredi Colli e S. Tommaso D'Aquino.

Fabroni, Giovanni

    1799   "Sur l'action chimique des différens métaux entr'eux, à la température commune de l'atmosphère, et sur l'explication de quelques phénomènes galvaniques." *Journal de Physique, de Chimie, d'Histoire naturelle et des Arts* 49:348–57.

Feyerabend, Paul

    1965   "Problems of Empiricism." In R. Colodny, ed., *Beyond the Edge of Certainty*, 145–260. Lanham, Md.: University Press of America.

    1975   *Against Method.* London: New Left Books.

Field, John, ed.

    1959   *Handbook of Physiology. Section 1: Neurophysiology.* Washington, D.C.: American Physiological Society.

Figuier, Louis

    1868   *Les merveilles de la science ou description populaire des inventions modernes.* Vol. 1. Paris: Furne, Jouvet et Cie.

Fontana, Felice

    1760   "Dissertation épistolaire adressée au R.P. Urbain Tosetti [Bologna, 23 May 1757]." French trans. in Haller 1760, 3:159–243.

    1767   *De irritabilitatis legibus nunc primum sancitis et de spirituum animalium in movendis musculis inefficacia.* Lucae (Lucca): Typis Iohannis Riccomini.

    1795   *Treatise on the Venom of the Viper.* 2 vols. London: John Cathel. Trans. of the French ed., *Traité sur le venin de la vipère.* 2 vols. Florence, 1781.

Fontana, Felice, and Caldani, Marc'Antonio

    1980   *Carteggio.* Ed. R. Mazzolini and G. Ongaro. Trento: Società di Studi Trentini di Scienze Storiche.

Fordyce, George
   1788   "The Croonian Lecture on Muscular Motion." *Philosophical Transactions*
          78:23–36.
Fowler, Richard
   1793   *Experiments and Observations Relative to the Influence Lately Discovered by M. Gal-*
          vani and Commonly Called Animal Electricity. Edinburgh: Duncan.
Franklin, Benjamin
   1773   *Oeuvres.* Vols. 1–2. French trans. by J. Barbeu Dubourg. Paris: Quillian.
   1941   *Benjamin Franklin's Experiments.* Ed. and with a critical and historical introduc-
          tion by I. Bernard Cohen. Cambridge, Mass.: Harvard University Press.
   1945   *Benjamin Franklin's Autobiographical Writings.* Selected and ed. by Carl Van
          Doren. New York: Viking Press.
   1970   *The Papers of Benjamin Franklin.* Ed. L. W. Labaree. Vol. 14, January 1 through
          December 31, 1767. New Haven and London: Yale University Press.
Fulton, John Farquhar
   1926   *Muscular Contraction and the Reflex Control of Movement.* Baltimore: Williams &
          Wilkins.
   1940   "Medicine and the Sciences." In *The Development of the Sciences*, 2d ser., ed.
          L. L. Woodruff. New Haven: Yale University Press.
Fulton, John Farquhar, and Cushing, Harvey
   1936   "A Bibliographical Study of the Galvani and the Aldini Writings on Animal
          Electricity." *Annals of Science* 1:239–68.
Galvani, Luigi
   1792   *De viribus electricitatis in motu musculari commentarius cum Joannis Aldini dissertatione*
          et notis. Mutinae (Modena): Apud Societatem Typographicam.
   1938   "La scoperta dell'elettricità animale nella corrispondenza inedita fra Luigi Gal-
          vani e Lazzaro Spallanzani, con due lettere di Mariano Fontana e Bartolomeo
          Ferrari." Ed. L. Barbieri. *Atti e Memorie della Deputazione di Storia Patria Emilia*
          e Romagna 3(1937–38):69–97.
   1966   *De Ossibus Lectiones Quattuor, nunc primum editae.* Ed. M. Pantaleoni and
          G. Galboli. Bononiae (Bologna): In Aedibus Compositori.
Geddes, Leslie A., and Hoff, Hebbel E.
   1971   "The Discovery of Bioelectricity and Current Electricity: The Galvani–Volta
          Controversy." *IEEE Spectrum* 8, no. 12:38–46.
Gherardi, Silvestro
   1841a "Rapporto sui MMSS del Galvani legati all'Accademia dal Prof. Giovanni
          Aldini." GO, 1–106.
   1841b "Frammenti inediti di Galvani." GO, 461–79.
Gill, Sydney
   1976   "A Voltaic Engima and a Possible Solution to It." *Annals of Science* 33:351–70.
Giulotto, Luigi
   1987   *La fisica a Pavia nell'800 e '900. Scritti di Luigi Giulotto.* Pavia: Università degli
          Studi di Pavia.
Gliozzi, Mario
   1935   "Giambatista Beccaria nella storia dell'elettricità." *Archeion* 17, no. 1:15–47.
   1937   *L'elettrologia fino al Volta.* 2 vols. Naples: Loffredo.

Gliozzi, Mario (*cont.*)

1962 "Fisici piemontesi del Settecento nel movimento filosofico del tempo." *Filosofia* 13:559–71.

1966 "Il Volta della seconda maniera." *Cultura e Scuola* 5, no. 17:235–39.

Gray, Stephen

1731 "Several Experiments Concerning Electricity." *Philosophical Transactions* 37: 18–44.

Griselini, Francesco

1748 "Lettera intorno l'Elettricità e alcune particolari esperienze della medesima [30 June 1747]." In Pivati 1748, 29–70.

Guericke, Otto von

1672 *Experimenta nova (ut vocantur) magdeburgica de vacuo spatio.* Amstelodami (Amsterdam): Apud J. Janssonium à Waesberge.

Guyot

1786 *Nouvelles récréations physiques et mathématiques.* 4 vols. Paris: Gueffier.

Haller, Albrecht von

1760 *Mémoires sur les parties sensibles et irritables du corps animal.* 3 vols. Lausanne: Sigismon d'Arnay.

1766 *Elementa physiologiae corporis humani.* Vol. 4. Lausanne: Sumptibus Francesci Grasset et Sociorum.

Heilbron, John L.

1966a "G. M. Bose: The Prime Mover in the Invention of the Leyden Jar?" *Isis* 57:264–67.

1966b "A propos de l'invention de la bouteille de Leyde." *Revue d'histoire des sciences* 19:133–42.

1971 "The Electrical Field before Faraday." In G. Cantor and J. Hodges, eds., *Conceptions of Ether*, 187–213. Cambridge, England: Cambridge University Press.

1977 "Volta's Path to the Battery." In G. Dubpernell and J. H. Westbrook, eds., *Proceedings of the Symposium on Selected Topics in the History of Electrochemistry*, vol. 78–6. Princeton, N.J.: The Electrochemical Society Inc.

1979 *Electricity in the 17th and 18th Centuries: A Study of Early Modern Physics.* Berkeley, Los Angeles, and London: University of California Press.

1987 "The Contributions of Bologna to Galvanism." Paper read at the "Universitates et Università" conference held at the Alma Mater Studiorum Saecularia Nona celebrations, Bologna, 16–21 November.

Hoff, Hebbel E.

1936 "Galvani and the Pre-Galvanian Electrophysiologists." *Annals of Science* 1:157–72.

Home, Roderick

1970 "Electricity and the Nervous Fluid." *Journal of the History of Biology* 3:235–51.

1972 "Franklin's Electrical Atmospheres." *British Journal for the History of Science* 6, no. 22:131–51.

1979 "Introduction" to Aepinus 1979, 1–224.

1981 *The Effluvial Theory of Electricity.* New York: Arno Press.

Humboldt, Alexander von

1799 *Expériences sur la galvanisme et en général sur l'irritation des fibres musculaires et nerveuses.* French trans. by J.-F.-N. Jadelot. Paris: Imprimerie Didot Jeune. 1st

German ed.: *Versüche über die gereizte Müskel- und Nervenfaser, nebst Vermuthungen über den chemischen Process des Lebens in der Thier- und Pflanzenwelt.* 2 vols. Posen: Decker; Berlin: H. A. Rottman, 1797.

Jallabert, Jean
  1748   *Expériences sur l'éléctricité avec quelques conjectures sur la cause de ses effets.* Geneva: Barillot & Fils.

Kipnis, Naum
  1987   "Luigi Galvani and the Debate on Animal Electricity, 1791–1800." *Annals of Science* 44:107–42.

Kuhn, Thomas
  1962   *The Structure of Scientific Revolutions.* Chicago: The University of Chicago Press. (2d ed., 1970).
  1977   "Objectivity, Value Judgment and Theory Choice." In *The Essential Tension,* 320–39. Chicago and London: The University of Chicago Press.
  1987   "What Are Scientific Revolutions?" In L. Krüger, L. J. Daston, and M. Heidelberger, eds., *The Probabilistic Revolution,* vol. 1, *Ideas in History.* Cambridge, Mass.: The MIT Press.

Laghi, Tommaso
  1756   "Epistola ad Caesareum Pozzi." In Fabri 1757b, 110–16.
  1757   "Sermo alter de sensitivitate atque irritabilitate halleriana." In Fabri 1757b, 326–44.

Lakatos, Imre
  1978   "Falsification and the Methodology of Scientific Research Programmes" (1970). In *Philosophical Papers,* 2 vols., 1:8–101. Cambridge, England: Cambridge University Press.

La Métherie, Jean-Claude de
  1802   "Discours préliminaire." *Journal de physique, de chimie, d'histoire naturelle et des arts* 56 (Nivôse Year XI [Dec. 1802-Jan.1803]):5–92.

Laudan, Larry
  1977   *Progress and Its Problems.* Berkeley: University of California Press.

Lemay, Joseph A. Leo
  1961   "Franklin and Kinnersley." *Isis* 52:575–81.

Le Monnier, Louis-Guillaume
  1746   "Recherches sur la communication de l'électricité." *MAS,* 447–64.

Mahon, Lord: *see* Stanhope, Charles

Mamiani, Maurizio
  1983   "Il galvanismo di Galvani." In Luigi Galvani, *Memorie sull'elettricità animale al celebre abate Lazzaro Spallanzani,* ed. M. Recenti, 23–29. Rome: Edizioni Theoria.

Mangin, Abbé
  1752   *Histoire générale et particulière de l'électricité.* Paris: Rollier.

Marat, Jean-Paul
  1782   *Recherches physiques sur l'électricité.* Paris: De l'Imprimerie de Clousier.
  1784   *Mémoire sur l'électricité animale.* Paris: L'Imprimerie de J. Lorry.

Marum, Martinus van
  1785   "Description d'une très grande machine électrique placée dans le Museum de Teyler, à Harlem, et des expériments faits par le moyen de cette machine."

Marum, Martinus van (*cont.*)

English trans. in Robert J. Forbes et al., eds., *Martinus van Marum: Life and Work*, 5:29–33. Haarlem: Hollandische Maatschappij der Wetenschappen, Tjeenk Willink, 1971.

Massardi, Francesco

1926 "Sull'importanza dei concetti fondamentali esposti dal Volta nel 1769 nella sua prima memoria scientifica 'De vi attractiva ignis electrici'." *Rendiconti del R. Istituto Lombardo di Scienze e Lettere*, 2d ser., 59:373–81.

Matteucci, Carlo

1842 "Deuxième Mémoire sur le courant électrique propre de la grenouille et sur celui des animaux à sang chaud." *Annales de chimie et de physique* 3(8):301–39.

1867 *La pila di Volta*. Florence: Stabilimento Civelli.

Mauduyt, Pierre-Jean-Etienne

1776 "Premier mémoire sur l'électricité considérée relativement à l'économie animale et à l'utilité dont elle peut être en Médecine." *Histoire et Mémoires de la Société Royale de Médecine* (1776), 461–513.

1784 *Mémoire sur les différentes manières d'administrer l'électricité et observations sur les effets qu'elles ont produits*. Paris: De l'Imprimerie Royale.

Mauro, Alexander

1969 "The Role of the Voltaic Pile in the Galvani-Volta Controversy Concerning Animal vs. Metallic Electricity." *Journal of the History of Medicine* (April), 140–50.

Mazzolini, Renato

1986 "Il contributo di Leopoldo Nobili all'elettrofisiologia." In G. Tarozzi, ed., *Leopoldo Nobili e la cultura scientifica del suo tempo*, 183–99. Bologna: Nuova Alfa Editoriale.

Mazzolini, Renato, and Ongaro, Giuseppe

1980 "Introduzione" to Fontana and M. A. Caldani 1980.

Medici, Michele

1845 "Elogio di Luigi Galvani." *Memorie della Società Medico-Chirurgica di Bologna*, Bologna, 4:1–29. Reprinted in *Compendio storico della scuola anatomica di Bologna*, 362–78. Bologna: Tipografia Governativa della Volpe e del Sassi, 1857.

Mellina, Luigi

1981 "L'elettroshock come taglio epistemologico nell'àmbito dei trattamenti somatici." In L. Del Pistoia and F. Bellato, eds., *Curare e ideologia nel curare in psichiatria*, 145–62. Lucca: Maria Pacini Fazzi Ed.

Mesini, Candido

1958 *Luigi Galvani*. Bologna: Tipografia S. Francesco.

1971 *Nuove ricerche galvaniane*. Bologna: Tamari Editori.

Moruzzi, Giuseppe

1964 "L'opera elettrofisiologica di Carlo Matteucci," *Physis* 6(2):101–40.

Musschenbroek, Pieter van

1768 *Introductio ad Philosophiam naturalem*. Editio Prima Italica. Patavii (Padua): Apud Joannem Manfré.

Newton, Isaac

1952 *Opticks: Or a Treatise of the Reflections, Refractions, Inflections and Colours of Light*. Based on the 4th ed., London 1730, with a foreword by Albert Einstein, an introduction by Sir Edmund Whittaker, a preface by I. Bernard Cohen, and an analytical table of contents prepared by Duane H.D. Roller. New York: Dover Publications.

Nicholson, William

1802   "Observations on the Preceding Memoir [= Volta 1802]." *A Journal of Natural Philosophy, Chemistry, and the Arts*, n.s., 1(1802): 142–44. French trans. in *Bibliothèque britannique*, reprinted in VO, 2:154n.

Nobili, Leopoldo

1834   *Memorie ed osservazioni edite ed inedite del cavaliere Leopoldo Nobili*. 2 vols. Florence: Davide Passigli e soci.

Nollet, Jean-Antoine

1746a   "Observations sur quelques nouveaux phénomènes d'électricité." *MAS*, 1–23.

1746b   *Essai sur l'électricité des corps*. Paris: Les Frères Guérin.

1747   *Saggio intorno all'elettricità de'corpi*. Italian trans. of Nollet 1746b. Venice: Presso Giambattista Pasquali.

1748a   "Des effets de la vertu électrique sur les corps organisés." *MAS*, 164–99.

1748b   *Leçons de physique expérimentale*. 6 vols. (vol. 4). Paris: Les Frères Guérin.

1749a   *Recherches sur les causes particulières des phénomènes électriques et sur les effets nuisibles ou avantageux qu'on peut en attendre*. Paris: Les Frères Guérin.

1749b   "Expériences et observations faites en différens endroits d'Italie." *MAS*, 444–88.

1774   *Lettres sur l'électricité*, new ed. Paris: Durand.

1984   "Correspondance Nollet-Jallabert." In Benguigui 1984.

Pantaleoni, Marina

1966   "Commentario" and "Note" in Galvani 1966.

Parrot, Georg Friedrich von

1829   "Lettre à MM. les Rédacteurs des *Annales de Chimie et de Physique*, sur les phénomènes de la pile voltaïque." *Annales de chimie et de physique* 42:45–66.

Pera, Marcello

1982   *Apologia del metodo*. Bari and Rome: Laterza.

1986–87   "Narcissus at the Pool: Scientific Method and the History of Science." *Organon* 22/23:79–98.

1987a   "The Rationality of Discovery: Galvani's Animal Electricity." In M. Pera and J. C. Pitt, eds., *Rational Changes in Science*, 177–201. Dordrecht, Boston, and Lancaster: D. Reidel.

1987b   "From Methodology to Dialectics: A Post-Cartesian Approach to Scientific Rationality." In A. Fine and P. Machamer, eds., *PSA 1986*, 2:359–74. East Lansing, Mich.: Philosophy of Science Association.

1988a   "Breaking the Link between Methodology and Rationality: A Plea for Rhetoric in Scientific Inquiry." In D. Batens and J. P. van Bendegem, eds., *Theory and Experiment: Recent Insights and New Perspectives*, 259–76. Dordrecht, Boston, and Lancaster: D. Reidel.

1988b   "Radical Theory Change and Observational Equivalence: The Galvani-Volta Controversy." In W. R. Shea, ed., *Revolutions in Science: Their Meaning and Relevance*, 133–56. Canton, Mass.: Science History Publications/U.S.A.

1989   "How Crucial Is a Crucial Experiment? Reflections on the Galvani-Volta

Pera, Marcello (*cont.*)

Controversy." In A. Baruzzi, C. Franzini, E. Lugaresi, and P. L. Parmegiani, eds., *From Luigi Galvani to Contemporary Neurobiology*, 19–37. Fidia Research Series, vol. 22. Padova: Liviana Press, and Berlin: Springer Verlag.

1991 "The Role and Value of Rhetoric in Science." In M. Pera and W. R. Shea, eds., *Persuading Science: The Art of Scientific Rhetoric*, 29–54. Canton, Mass.: Science History Publications/U.S.A.

Pivati, Gianfrancesco

1746 *Nuovo Dizionario Scientifico e curioso sacro-profano*. 10 vols. Venice: Presso Benedetto Milocco, 1746–51.

1747 *Lettera della elettricità medica al celebre Signore Francesco Maria Zanotti* (29 July 1747). Lucca.

1748 *Lettere sopra l'elettricità principalmente per quanto spetta alla medicina*. Venice: Appresso Simone Occhi.

1749 *Riflessioni fisiche sopra la medicina elettrica*. Venice: Presso Benedetto Milocco.

Polvani, Giovanni

1942 *Alessandro Volta*. Pisa: Domus Galileana.

Popper, Sir Karl

1972 *Objective Knowledge: An Evolutionary Approach*. Oxford: Clarendon Press.

Priestley, Joseph

1775a *Experiments and Observations on Different Kinds of Air* (1774). 2d ed., 2 vols. London: For J. Johnson.

1775b *The History and Present State of Electricity* (1767). 3d ed., 2 vols. London: For C. Bathurst, T. Lowndes, etc. Reprint: New York: Johnson Reprint Co., 1966.

Pupilli, Giulio Cesare

1956 "Luigi Galvani." *Studi e Memorie per la storia dell'Università di Bologna*, n.s., 1:445–59.

Pupilli, Giulio Cesare, and Fadiga, Ettore

1963 "The Origins of Electrophysiology." *Journal of World History* 7(2):547–89.

Recenti, Mauro

1983 "Introduzione" to Luigi Galvani, *Memorie sull'elettricità animale al celebre abate Lazzaro Spallanzani*, ed. M. Recenti, 5–21. Rome: Edizioni Theoria.

Robinson, Bryan

1734 *A Treatise of the Animal Oeconomy* (1732), 2d ed. Dublin: George Ewing and William Smith.

Robison, John

1793 "Letter to Mr. Fowler [Edinburgh, 28 May 1793]." In Fowler 1793, 169–76.

Rothschuh, Karl E.

1960a "Von der Idee bis zum Nachweis der tierischen Elektrizität." *Sudhoffs Archiv* 44:25–44.

1960b *Alexander von Humboldt et la découverte de l'électricité animale*. Paris: Edition du Palais de la Découverte, ser. D, no. 72.

1963 "Die neurophysiologischen Beiträge von Galvani und Volta." In Belloni 1963, 117–30.

1973 *History of Physiology*. New York: Krieger.

Rowbottom, Margaret, and Susskind, Charles
   1984   *Electricity and Medicine: History of Their Interaction.* San Francisco: San Francisco Press.
S: *see* Spallanzani 1792
Sguario, Eusebio
   1746   *Dell'elettricismo.* Venice: Presso Gio. Battista Recurti.
Sigaud de la Fond, Joseph-Aignan
   1785   *Précis historique et expérimental des phénomènes électriques depuis l'origine de cette découverte jusqu'à ce jour.* Paris: Rue et Hôtel Serpente.
Sirol, Marc
   1939   *Galvani et le galvanisme. L'électricité animale.* Paris: Vigot Frères.
Spallanzani, Lazzaro
   1792   "Transunto della Dissertazione del Sig. Dott. Luigi Galvani P. Prof. nell'Università di Bologna Sulle forze dell'Elettricità nei moti muscolari." *Opuscoli scelti sulle scienze e sulle arti* (1792), 15:113–41. Signed "S".
   1964   *Epistolario.* Ed. Prof. Benedetto Biagi. Florence: Sansoni Antiquariato.
   1976   "Carteggio di Lazzaro Spallanzani con Luigi Galvani e Giovanni Aldini," ed. P. di Pietro. *Atti e Memorie della Accademia Nazionale di Scienze, Lettere e Arti di Modena,* 6th ser., 18:115–46.
   1977   "Nuovo contributo al carteggio tra Lazzaro Spallanzani e Luigi Galvani," ed. P. di Pietro. *Atti e Memorie della Accademia Nazionale di Scienze, Lettere e Arti di Modena,* 6th ser., 19:167–72.
Stanhope, Charles (Lord Mahon)
   1779   *Principles of Electricity.* London: P. Elmsly.
Sue, Pierre
   1802   *Histoire du Galvanisme et analyse des différens ouvrages publiés sur cette découverte depuis son origine, jusqu'à ce jour.* 4 vols., 1802–5. Paris: Bernard.
Sulzer, Johann Georg
   1767   *Nouvelle théorie des plaisirs avec des réflexions sur l'origine du plaisir par Mr. Kaestner.* N.p.
Sutton, Geoffrey
   1981   "Electrical Medicine and Mesmerism." *Isis* 72:375–92.
Symmer, Robert
   1759   "New Experiments and Observations Concerning Electricity." *Philosophical Transactions* 61:340–89.
Tabarroni, Giorgio
   1966   "La torre dell'Università di Bologna e l'elettricità atmospherica." *Coelum* 34:5–6.
   1971a   "Dall'elettricismo all'elettromagnetismo." In A. Pasquinelli and G. Tabarroni, eds., *Le teorie scientifiche dalla fine del Settecento al 1860,* 95–220. Vol. 21 of M. F. Sciacca, ed., *Grande Antologia Filosofica.* Milan: Marzorati.
   1971b   "Galvani, Aldini e la corrente elettrica." *Annuario 1969–70 dell'Istituto Tecnico Industriale Aldini-Valeriani,* 43–54. Bologna.
Taglini, Carlo
   1747   *Lettere scientifiche sopra vari dilettevoli argomenti di fisica.* Florence: Nella Stamperia all'Insegna d'Apollo.

Torlais, Jean
  1954 *Un physicien au siècle des lumières, l'abbé Nollet.* Paris: SIPUCO.
  1956 "Une grande controverse scientifique au XVIIIe siècle. L'abbé Nollet et Benjamin Franklin." *Revue d'histoire des sciences* 9:339–49.
  1969 "Qui a inventé la bouteille de Leyde?" *Revue d'histoire des sciences* 16:211–19.
Valli, Eusebio
  1793a *Experiments on Animal Electricity with their Application to Physiology and Some Pathological and Medical Observations.* London: J. Johnson.
  1793b "IXe Lettre sur l'Electricité animale." *Observations sur la physique* 42:74–75.
  1794 *Lettera XI sull'elettricità animale.* Mantua: Nella Stamperia di Giuseppe Braglia.
Van Doren, Carl
  1938 *Benjamin Franklin.* New York: Viking Press.
Vassalli-Eandi, Antonio Maria
  1803a "Expériences et observations sur le fluide de l'électro-moteur de Volta." *Mémoires de l'Académie des Sciences de Turin* 11:1–34.
  1803b "Saggio sopra il fluido galvanico." *Memorie di matematica e fisica della Società Italiana delle Scienze* 10:733–765.
Veratti, Giuseppe
  1748 *Osservazioni fisico-mediche intorno alla Elettricità.* Bologna: Nella Stamperia di Lelio della Volpe.
  n.d. *Osservazione fatta in Bologna l'anno MDCCLII de i fenomeni elettrici nuovamente scoperti in America e confermati a Parigi.* Bologna: Nella Stamperia di Lelio della Volpe.
Vivenzio, Giovanni
  1784 "Istoria dell'elettricità medica." Preface to Cavallo 1784, 5–66.
Volpati, Carlo
  1927 *Alessandro Volta nella gloria e nell'intimità.* Milan: Treves.
Volta, Alessandro
  1802 "Letter of Professor Volta to J. C. Delamethrie [sic], on the Galvanic Phenomena [Paris, 18 Vendémiaire Year XI (10 Oct. 1801)]." *A Journal of Natural Philosophy, Chemistry, and the Arts,* n.s., 1(1802):135–42. French original published in *Journal de physique* 53 (1801) and reprinted in VO, 2:37–43.
  1952 "On the Electricity excited by the mere Contact of conducting Substances of different Kinds." Facsimile of *The Philosophical Magazine* (Sept. 1800) translation (probably by William Nicholson) in Dibner 1952, 42–49. French original in VO, 1:563–87.
Volta, Alessandro, Jr.
  1900 *Alessandro Volta e il suo tempo.* Carrara: Bertarelli.
Volta, Zanino
  1879 *Alessandro Volta. Della giovinezza,* pt. 1. Milan: Civelli.
Winkler, Johann Heinrich ("John Henry")
  1746 "An extract of a Letter Concerning the Effects of Electricity upon Himself and his Wife." *Philosophical Transactions* 44:211–12.
Wolf, Abraham
  1961 *A History of Science, Technology, & Philosophy in the 18th Century.* 2d ed., rev. by Douglas McKie. 2 vols. New York: Harper.

Name Index